Charles Sumner Dolley

The Technology of Bacteria Investigation

Charles Sumner Dolley

The Technology of Bacteria Investigation

ISBN/EAN: 9783743687264

Printed in Europe, USA, Canada, Australia, Japan

Cover: Foto ©berggeist007 / pixelio.de

More available books at **www.hansebooks.com**

THE

TECHNOLOGY

OF

BACTERIA INVESTIGATION

EXPLICIT DIRECTIONS FOR THE
STUDY OF BACTERIA

THEIR

CULTURE, STAINING, MOUNTING, ETC., ACCORDING TO
THE METHODS EMPLOYED BY THE MOST
EMINENT INVESTIGATORS

BY

CHARLES S. DOLLEY, M.D.

Boston
S. E. CASSINO AND COMPANY
1885

PREFACE.

The following pages are presented with the hope that they will stimulate careful study of the Schizomycetes by American investigators, and assist them in adding their share, as they have done in other branches of scientific research, to the mass of facts concerning Bacteria; which, with two or three exceptions, have been thus far brought to light by European students.

C. S. D.

Naples Zoological Station, April 3, 1885.

TABLE OF CONTENTS.

Part I.

GENERAL DIRECTIONS 1

I. MICROSCOPICAL PREPARATIONS 2
The Study of Living Forms 3
Schnetzler's Method of obtaining Bacteria from the Air . 7
Brautlecht's Method of obtaining Bacteria from the Air . 8
Miquel's Method of obtaining Bacteria from the Air . . 8
Miquel's Second Method for the Analysis of Rain . . . 9
Miquel's Method for the Analysis of Drinking Water . . 10
Miquel's Method for ascertaining the Hourly Variations of Aerial Bacteria 11
Hesse's Method for the Quantitative Estimation of Microorganisms contained in the Air 15
Marchand's Method for the Examination of Microbes or Corpuscles held in Suspension in Water 16
Literature of Bacteria contained in Air, Water, or Earth . 17
Naegeli's Method for the Isolation of Particular Bacterial Forms 22
Buchner's and Robert's Method for obtaining Bacillus Subtilis 23
Tiegel's Method for proving the Existence of Bacteria in the Tissues of Healthy, Living Organisms 23
Burdon-Sanderson's Method for proving the Existence of Bacteria in the Tissues of Healthy Living Organisms . . 24
Chiene and Ewart's Method for proving the Existence of Bacteria in the Tissues of Healthy Living Organisms . . 25
Nencki and Giacosi's Methods for proving the Existence of Bacteria in Healthy Living Organisms 25
Literature on Bacteria in Healthy Living Organisms . . 27
To obtain Blood for Microscopical Examination . . . 32
Moss's Method for obtaining Blood for Investigation . . 32
To obtain Fluid or Serum from a freshly extracted Organ . 33

TABLE OF CONTENTS.

The Study of Microscopical Preparations of Fixed and Hardened Bacteria	34
The Use of Staining Fluids	35
Hazelwood's General Blue Stain	37
Blanchard's Method of Staining "Proliferous Membranes,"	37
"T. C.'s" Method	38
Cover-glass Preparations	39
General Method for Staining Cover-glass Preparations	41
Koch's Method for Staining Cover-glass Preparations	41
Gram's Method for Staining Cover-glass Preparations	42
The Preparation of Bacteria in and upon Firm Substances	43
Weigert's Method	44
Schütz's Method	45
Babe's Method	45
Subbotin's Combination of Staining Fluids	46
Mounting	46
Preparation of Bacteria for Photographing	48
Sternberg's Method	48
Kaschka's Method	48
Dufrenne's Method	49
Literature of the Preparation and Investigation of Bacteria in General	51
II. CULTURE EXPERIMENTS	56
Fluid Culture Media	59
Pasteur's Sterilization Method	60
Buchner's Sterilization Method	61
Tyndall's Sterilization Method by Discontinuous Heating	61
Miquel's Sterilization Method without Heat	61
Gautier's Method of Sterilization without Heat	62
Koch, Gaffky, and Loeffler's Steam Sterilizing Cylinder	63
Solid or Gelatine Culture Media	64
Koch's Serum Culture Medium	66
Culture Vessels	67
Salmon's Culture Tube	67
Sternberg's Culture Flasks	70
To introduce the Sterilized Culture Fluid into the Sterilized Culture Vessel	71
To sow the Microbes in the Culture Medium	72
To carry on the Cultures	73
Van Tieghem and Lemonnier's, and Miquel's Methods for the Cultivation of Bacteria upon the Slide	73

TABLE OF CONTENTS. vii

Salmonsen's Method for Pure Cultures 75
Literature on Culture Methods 76

III. VACCINATION OR INOCULATION EXPERIMENTS . . 78
Literature of Inoculation Methods 81

IV. BIOLOGICAL ANALYSIS 82
Engelmann's Method for Studying the Effect of Light on
   Certain Bacteria 85
Engelmann's Method for the Determination of a Chromophyll
   Assimilating Power in any given Bacteria 86
Duclaux's Method for Studying the Effect of Sunlight upon
   the Germs of Bacteria 87
Pictet and Yung's Method for Ascertaining the Action of Cold
   upon Microbes 88
Literature on the Biology of Bacteria 88

Part II.

SPECIAL METHODS FOR INVESTIGATING PATHOGENIC BACTERIA 93

ANTHRAX 99
Toussaint's Method 99
Pasteur's Method 100
Chaveau's Methods 103
Chamberland and Roux's Method 106
Koch, Gaffky, and Loeffler's Method 107
Weigert's Staining Method 109
Feltz's Method of Ascertaining the Rôle of Earthworms in
   the Propagation of Anthrax 109
Literature 109

CHOLERA 117
Koch's Methods 118
Nicati and Rietsch's Method 126
Bochefontaine's Method 126
Literature 127

GLANDERS 131
Loeffler and Shutz's Method 131
Literature 132

TABLE OF CONTENTS.

Hog Cholera	133
Literature	134
Hydrophobia	135
Gibier's Method of Attenuation	135
Pasteur's Method of Inoculation	135
Babe's Staining Method	137
Literature	139
Leprosy	140
Babe's Staining Method	141
Baumgarten's Staining Method	141
Neisser's Methods	142
Literature	143
Malaria	145
Klebs and Tommassi-Crudelli's Method for Collecting Malaria Germs	145
Richard's Method for the Study of the Blood	147
Literature	147
Malignant Œdema	150
Pasteur's Method	150
Literature	150
Septicæmia of Domestic Mouse	151
Septicæmia of Rabbits	152
Literature	153
Syphilis	153
Birsch-Hirschfeld's Method	153
Kleb's Method	153
Literature	154
Tuberculosis	155
Koch's Method	156
Ehrlich's Methods	157
Ziehl's Method	158
Balmer and Fraenzel's Methods	159
Rindfleish's Method	159
Orth's Method	160
Lichtheim's Method	161
Petri's Method	161
Gibb's Methods	163
Baumgarten's Methods	164

TABLE OF CONTENTS.

Weigert's Method 166
Fraenkel's Method 166
Pfuhl-Petri's Method 167
Senkewitch's Method 168
Käätzer's Method 168
Long's Method 169
Peter's Method 169
Veraguths' Method 170
Coze and Simon's Staining Methods 171
Déjérine's for Study of Concrements from Lungs . . 171
Giboux's Method of Testing the Inoculability of Tuberculosis through Respiration 172
Ermengen's Staining Method 172
Brun's Staining Method 173
Burrill's Staining Method 174
Hartzell's Staining Method 174
Quinlan's Staining Method 175
Negri's Method for Staining the Spores 176
Reinstadler's Method 178
Celli and Guarnieri's Method for Ascertaining whether the Microbes of Tuberculosis are thrown into the Air by Expiration 179
Literature 181

TYPHOID FEVER 193
Letzerich's Method 193
Rindfleish's Method for Contaminated Well-water . . 195

WHOOPING COUGH 197
Burger's Method 197
Literature 197

CONCRETIONS OF THE LACHRYMAL DUCTS . . . 197
Literature 197

DENTAL CARIES 198
Leber's Method 198
Miller's Method 199
Literature 199

AREA CELSII 200
Buchner's Method of Study 200
Von Sehlen's Method 200
Literature 201

TABLE OF CONTENTS.

CHICKEN CHOLERA 201
Barthelmy's Method of Study 202
Pasteur's Method 202
Literature 203
DIPHTHERIA 204
Loeffler's Method 204
Literature 205
ERYSIPELAS 207
Fehleisen's Methods 207
Literature 207
FURONCLE 208
Pasteur's Method 209
Literature 209
GONORRHŒA 210
Neisser's Methods 210
Literature 211
OSTEOMYELITIS 213
Krause's Methods 213
Literature 214
PNEUMONIA-CROUPOSA 215
Friedlaender's Methods 215
Literature 217
RECURRENT FEVER 220
Koch's Method 220
Friedlaender's Method of Obtaining the *Spirilla* . . . 220
Literature 221
YELLOW FEVER 222
Friere's Methods 222
Literature 222
MISCELLANEOUS PATHOGENIC BACTERIA, WITH LITERATURE 223
Contagious Septicæmia 223
Rheumatic Arthritis 225
Endocarditis Ulcerosa 225
Infectious Wound Diseases 226
Variola 227
Influenza Epidemica 229
Meningitis 229
Measles 230

Phlegmon 230
Scarlet Fever 230
Molluscum Contagiosum 231
Dilatation and other Diseases of the Stomach 231
Anæmia Perniciosa 232
Ulcerative Stomatitis in the Calf 232
Cattle Plague 232
Acute Yellow Atropy of the Liver 232
Dysentery 233
Ozæna 233
Foot Rot 233
Verruga Peruana 233
Diabetes 233
Chyluria 234
Fœtid Feet 234
Malleus humidus 234
Pyæmia 234
Diseases of Insects 234

BACTERIA IN PLANT TISSUES 235
Ducleaux's Method of Studying the Effect of Microbes upon
   Germination 235
Ralph's Method for Demonstrating the Presence of Microbes
   in the Cells of Aquatic Plants 235
Literature 236

Part III.

FORMULARY 238
Bergmann's or Bucholz's Fluid 238
Bismarck-Brown 238
Glycerine Aniline-Brown 239
Aniline Oil 239
Aniline Water 240
Aniline-Yellow 240
Acidulated Fluids for Decolorizing 240
Ether 241
Alcohol 241
Canada Balsam 241
Chrysoidin 241

TABLE OF CONTENTS.

Cement for Glycerine Mounts 242
Culture Fluids 242
Eosin Hæmatoxylin 243
Eosin 243
Fuchsin 244
Fuchsin-Aniline-Oil 245
Gentian-Violet 245
Kleb's Glycerine-Jelly 246
Glycerine 246
Boehmer's Hæmatoxylin 247
Methyl-Blue 247
Methyl-Green and Malachit-Green 248
Methyl-Violet 248
Magenta 249
Nigrosin or Aniline-Black 249
Orseille 250
Oil of Cloves 250
Osmic Acid 250
Peptone Solution 251
Picrocarmine 251
Picrocarminate of Ammonia 252
Acetate of Potash for Mounting 252
Caustic Potash Solutions 253
Rosaniline Chlorohydrate 253
Acid Fuchsin 253
Vesuvin 254
Plaster of Paris Filters 254
Cleaning Fluids 254
Soap Imbedding Mass 255
Celloidin Imbedding Mass 256
Brun's Mounting Medium 258
Gautier's Red Lead Cement 259
Nutritive Paper 259
Solid Culture-Media 260
Lichen Jelly 260
Hydrocele Fluid 261
Blood Serum 262
Agar Agar 262
Gelatine 263

BACTERIA INVESTIGATION.

PART I.

GENERAL DIRECTIONS.

IN the study of the Schizomycetes it will be found of advantage to use slides of the whitest possible glass, of from 1 to 2 mm., and cover-glasses of from 0.10 to 0.18 mm. in thickness. It is of the highest importance that these, as well as all other implements, such as tubes, pincers, needles, wires, etc., should be made perfectly clean immediately before their use. Metallic instruments can be best sterilized by heating in the flame of a spirit lamp or Bunsen burner. Slides and cover-glasses should be kept in a cleaning fluid [F. 40], and washed and wiped well before using. Other glassware may be cleaned by the use of this same fluid, or by passing it through concentrated hydrochloric acid and afterwards washing it in hot alcohol.

All staining-fluids should be filtered through good Swedish filter-paper previous to use, and if much importance is attached to the investigation, the filtrate should be examined microscopically.

The presence and variety of Bacteria may, by the trained eye, be diagnosed with objectives of medium power [230 : 540 diam.], but in the study of pathogenic micro-organisms a good homogeneous immersion objective and an Abbe's condensor are considered essentials. The best light is that obtained from white clouds, or sunlight reflected from a white wall; next to this comes the light obtained from good kerosene, which is improved by the addition of a small piece of camphor. To protect the eyes from the glare of the light, Engelmann regards a green or greenish-blue glass placed between the object and the light as preferable to the cobalt-blue glass ordinarily used, or to a blue lamp chimney, while Flesch considers a pale yellow glass better than the green. In no case should the colored glass be placed between the object and the eye. In searching for Bacteria in any organ, it is better first to study such fluids, serous or other, as may be obtained from the organ, and then the organ itself. The methods used for the study of Bacteria may be grouped under four heads, *i. e.*, I. Microscopical preparations; II. Culture experiments; III. Vaccination or Inoculation experiments; IV. Biological analysis.

I. Microscopical Preparations.

In this method the microbes[1] are to be distinguished from similar inorganic particles, either by the characteristic form of single individuals, by the arrangement of these into characteristic groups or colonies, or by their relation to chemical reagents. It must be borne in mind that, with few exceptions, all Bacteria at different stages of their life history exhibit different forms. The spores or germs of all Bacteria are round, micrococcus-like bodies, from which develop rod-like forms, spiral or straight filamentous forms, and eventually the forms typical of the species. We may study these in microscopical preparations, either alive in some transparent nourishing medium, or after killing and fixing them. The first of these methods, *i.e.*, *The Study of Living Forms*, is not best adapted for persons unacquainted with the various bacterial types. These should first be

---

[1] The word "microbe" was introduced by Mr. Chas. Sédillot as follows : —

"Les noms de ces organismes sont très nombreux et devront être definis et, en partie, réformés. Le mot *microbe* (Μιχρός, petit, Βίος, vie) ayant l'avantage d'être plus court et d'une signification plus générale, et mon illustre ami, M. Littré, le linguiste de France le plus compétent, l'ayant approuvé, nous l'adoptons, sans néanmoins renoncer à ceux en usage, pour la désignation de variétés plus particulièrement etudiées." Vid. "De l'influence des découvertes de M. Pasteur sur les progrès de la chirurgie." Compt. Rend. T. 86, p. 634.

rendered familiar by the study of fixed and stained preparations. When this has been accomplished, the living microbes may be studied for the purpose of learning their peculiarities of movement, grouping, behavior towards various reagents, and the finer details of structure lost by drying, etc., by placing upon a properly cleaned cover-glass a drop of the fluid containing them, and inverting it over a shallow cell or the cavity of a hollowed-out slide, or by a thin layer of the fluid spread out and covered upon an ordinary slide. If too large a number of microbes are in the field at once, producing a confused picture, the liquid may be thinned by the addition of a drop of pure filtered water, common salt solution, or peptone solution [F. 31]. The investigator is not obliged to search far for specimens of Bacteria upon which to institute comparative studies. The atmosphere of the laboratory, of the home, of city and country, as has been shown by Pasteur, Miquel, and others, is a rarely failing source of living germs, if not of developed microbes. Flowing and stagnant water, both fresh and salt, affords respectively peculiar forms; the so-called "filth" Bacteria may be sought in infusions of decaying animal or vegetable matter; those forms which act as special ferments will, of course, be found in fluids undergoing butyric, lactic, or other fermentation, as the case may be.

The various fluids of living animal bodies in health and disease afford an abundant harvest of microbes illustrating nearly every group of the Schizomycetes.

Among invertebrate animals Metschnikoff found in the serous fluids large numbers of Bacteria, which were being eagerly pursued and devoured by the amæboid cells of the animal. In fishes Richtet has shown that the blood and lymph contain Bacteria. The secretions of healthy human beings have been shown by Sternberg and others to abound in a variety of forms; thus the saliva has repeatedly been demonstrated to be rich in micro-organisms of great variety, long *Leptothrix* threads of different thickness, round *Micrococci* of varying size, in heaps and chains; often *Bacilli* and beautiful *Spirochæte*. All these, together with numerous specimens of the lower orders of Fungi, may be examined by preparing as above, or after being fixed and hardened, as directed further on. The vaginal secretions, like those of the mouth, furnish a nidus for microbes of great variety. In the stomach the ordinary bacterial forms are found as a rule, only in small numbers, owing probably to the acidity of the digestive juices. The intestine, on the other hand, is rich in Bacteria, *Micrococci* and *Bacilli* of all sizes and shapes; among the latter, at times, the ileum and colon contain specimens of *B. amylobacter*, the source of butyric acid fermentation. Besides the above, which appear in their fully-developed condition, Billroth, Nencki, and others have proven conclusively that the tissues of living, healthy animals contain myriads of bacterial germs or spores, while pathologists, in their study of the diseased body, have indicated the presence of many peculiar forms, some of which are regarded as the specific agents of infectious disease. Thus the lachrymal ducts and the sweat are at times inhabited by two very different salt-loving forms; the vaginal and urethral secretions, as well of those of the eyes, become at times the home of the gonorrhœal microbe; the stomach, in cases of catarrh, dilatation, and cancer, is often occupied by *Sarcina*, which also, along with other Bacteria, makes its home in the vermiform appendix of our domestic fowls. The intestine of the typhoid fever or cholera patient furnishes the *Bacilli* peculiar to these diseases. The lymph in syphilis, measles, erysipelas, etc., shows other forms, and the blood of diphtheria patients and those suffering from recurrent fever still others. Others take up their abode in various organs, as has been shown in leprosy and tuberculosis; the pus of abscesses — tubercular, furuncular, and otherwise — present *Micrococci* and *Bacilli* in abundance, while decayed teeth are already filled with an amalgam of Bacteria.

The enumeration of pathogenic Bacteria is, however, at this point out of place, it being only intended here to indicate the success which surely awaits the seeker after microbes in healthy and diseased animals. All or any of these forms may be studied as microscopical preparations, alive, by the simple methods we have indicated, or they may be subjected, while under the cover-glass, to a number of manipulations, which will be fully described under "*Biological Analysis.*"

BACTERIA OF THE AIR, WATER, AND EARTH.

Although Bacteria, as we have seen, are almost omnipresent, experience has shown that certain methods are to be preferred in bringing them into position for microscopic study. Thus it has been found that in the study of *Aerial Bacteria* the ordinary methods employed for collecting the dust of the atmosphere are of slight use, since it is difficult to distinguish the germs among a large number of inanimate granules and in a miscellaneous mixture of debris.

Schnetzler points out that we have at our command, for the study of aerial germs, a small apparatus traversed by about 8,000 cubic centimeters of air per minute, which may inform us as to these floating germs. This is no other than the nasal cavity, on the mucous surface of which air particles are deposited. To observe these he advises injecting the nose with distilled water (completely sterilized) by means of a glass syringe previously calcined. The liquid thus obtained is to be subjected to examination by any or all the methods employed for the microscope, or culture.

Brautlecht's Method, employed in the course of a research on the presence of Bacteria in the effluvia and vapors of fever districts, was to mix baked sand, gritty earth, and tolerably loamy garden mould, with liquids containing Bacteria, and then covering the mixture with a bell-glass, using all ordinary precautions against external contamination. After a few hours he found in the vapors condensed under the bell-glass a large number of microbes, of the same form invariably as those contained in the liquid used.

Miquel's Method of obtaining Bacteria from the air is by the use of rain, which contains more microbes than the water condensed artificially from the atmosphere by means of " aeroscopes," by ventilators or other appliances. He employs an apparatus which he calls an " Udometer." An iron rod is fastened horizontally to the upper portion of a wooden post set in the ground at a distance from any trees or habitation, where it will receive the first drops of rain. Two L-shaped arms, each terminating at one end in a ring and arranged at the other to slide upon the stationary rod and fasten with a thumbscrew, like the ordinary filter stands of the chemist, are provided. In the ring of one of these is placed, after being thoroughly sterilized at a high temperature, a nickel- or silver-plated funnel. In the second ring, which is arranged to hang directly below the first, is placed a platinum crucible, likewise sterilized, and for which a close-fitting cover is provided. By means of this apparatus, rain may be collected at the beginning, in the midst, or at the latter part of a storm, although the first rain contains the most bacterial germs. The cover being adjusted to avoid contamination during transport, the water is taken to the laboratory, where the Bacteria or their spores, which have been washed from the air, may be examined as microscopical preparations or by any of the other methods of study.

Miquel's Second Method for the analysis of rain. For the analysis of rain, Dr. Miquel has invented a special apparatus, — his "Udobactériemètre," — consisting, in brief, of a glass shade with its neck fitted with a stopper, through which is passed the long stem of a metal funnel that delivers, by droplets, the rain caught, and projects them upon the nutritive paper [F. 45] arranged in a very wide truncated cone, which is rotated at some speed by clockwork beneath. The droplets are carried by the rotation a little way along the sterilized paper cone, moistening the lichen surface in their course at different points, — no troublesome liquefaction taking place, as with ordinary gelatine.

The cone of paper, at the end of the experiment, is placed in the incubating stove with its damp atmosphere, and affords excellent results.

Miquel's Method for the analysis of drinking water. Now that analyses of drinking water are frequent, Dr. Miquel offers the following method, remarkable for the simplicity of its manipulations. In a small precipitate glass with a foot, and closed by an emery-ground tabulated cap, like Pasteur's flasks, a little wetted cotton is placed. Through the tube is passed a platinum wire, crooked at the end to support a band of the nutritive paper [F. 45] covered on both sides, about 3 cm. wide and 8 long, equivalent to about half a decimetre square surface; all is sterilized at 110° C. and then weighed to a milligramme. The mouth is now uncapped, the paper plunged for one or two minutes into water to be analyzed for the different organisms, then introduced into the *éprouvette*, and reweighed. If the water is impure, it is not long before small spots show the deposition and growth of the microbes, which spots can be counted; and, knowing the weight of water, the impurity can be rated in terms of comparison according to the time of immersion, — the increase in weight sometimes exceeding 2 gr. The various Bacteria, in their development on the nutritive paper, offer different appearances even to the naked eye. Pure cultures can thus also be obtained. In any case, when the germs are fully developed, the paper is again dried at 30–40° C., by which the colonies are fixed, and they can be preserved, photographed, or revived if the paper is varnished with a solution of gum. If a resin varnish is used, the microbes cannot be revived.

Miquel's Method for ascertaining the hourly variations of aerial Bacteria. Dr. Miquel had by previous methods shown the influence of rain, of aridity, of humidity, and of the force and direction of the wind, upon the quantity of germs held in suspension in the air; the annual, monthly, weekly, and daily variations in the quantity of these microscopic organisms. One more interesting point remained to be seen; *i. e.*, whether, in the same day,— the meteorological conditions remaining sufficiently fixed, — the numbers of Bacteria vary from hour to hour, like, for example, the thermometric pressure, the temperature, etc. Experiments described in the Annuaire de Montsouris firmly established the fact that at noon the number of Bacteria were twenty times less than at eight o'clock in the evening. It was, however, important to know whether these fluctuations occurred with regularity, and whether they could be detected every day at the same hour, providing, of course, they were not interrupted by sudden and unexpected meteoric phenomena. Again, experiments demonstrated that, under normal conditions, the variation of aerial Bacteria is effected with regularity. At eight o'clock in the morning, the number of atmospheric Bacteria is always high; after this time, it decreases until noon. From noon to one o'clock it is at a remarkable minimum, after which the numbers gradually increase again until eight in the evening. Experiments made at night gave identical results. At ten and eleven in the evening the air is very impure; from one to three in the morning it has purified itself considerably, to again have the number of microbes increase in the forenoon. The law concerning diurnal variation is true for the entire season. Diverse directions of the wind do not modify it, providing the directions remain constant. As to the force of the wind, one is astonished to find it without effect upon the phenomena of periodical increase and decrease. Contrary to the opinion held by many, the air is less pure morning and evening than at mid-day. In the present state of our knowledge, it is difficult to discern the cause of these regular variations. Dr. Miquel thinks, however, that the oblique currents determined on the surface of the earth by the heating and cooling of the soil have great effect in these phenomena. Those winds which sweep along the surface of the earth naturally charge themselves with a larger quantity of germs than those which arrive at the point of observation at an angle of incidence of 80° to 70°. Dr. Miquel regards the variable obliquity of the atmospheric currents as the cause of the increase and decrease of microbes. Unfortunately meteorology is dumb as to the nature, frequence, force, and periodicity of these oblique movements of the atmosphere. An anemometer should be invented having the faculty of measuring the inclination of the wind. Dr. Miquel has constructed a register of atmospheric Bacteria, which, if not capable of furnishing the exact number of germs, is capable of showing very closely their variations. The hourly statistics of Bacteria, by the process of enumeration employed heretofore, to be sufficiently exact, required 600–800 preparations of beef broth in twenty-four hours, besides five persons to carry on the experiments uninterruptedly day and night.

In confiding to a faithful register a work so expensive and fatiguing, a step in advance is realized which may be compared to the progress made in meteorology and physiology in the invention of thermographs, sphygmographs, etc. Some two years ago, Dr. Miquel described a Mucedinograph or Sporograph, which actually figured at the International Exposition of Hygiene at London, in the exceedingly instructive rooms at South Kensington, and by the side of which was to be seen a Bacteriometer which is far from being so simple an affair, and of which the following is a description. *Apparatus.*—Dr. Miquel adapted to a clockwork made by MM. Richard frères, a plated copper cylinder, or, better still, a cylinder of ebonite, upon the exterior of which was placed a band of *nutritive paper* [F. 45.] 10 cc. wide and 60 cc. long. The whole is placed under a large bell-glass having a tubular opening through the knob on its summit (through which aspiration is carried on). The borders of the bell-glass are placed in a groove made in a plate and filled with mercury. A vertical slit in the upright portion of the bell-glass allows the projection of a very fine jet of air upon the band of nutritive paper, which makes one revolution in twenty-four hours, and receives the microbes in the atmospheric dust at every instant of the day. *Course of the experiment.* 1st. The ebonite cylinder furnished with its nutritive paper is sterilized by exposure to a temperature of 110° C. for one hour. 2d. The cylinder is then placed upon the clockwork, and beneath the bell-glass, which is coated with vaseline. The time of starting is noted and a constant current of air is determined, varying from 30 to 60 litres per hour, according to circumstances. Twenty-four hours afterwards the aspiration is suspended, and the results noted by further treatment of the paper. 3d. Remove the ebonite cylinder with its band of paper, and place it in a second vaselined bell-glass, where the Bacteria may incubate and develop protected from all dust. *Precautions.* Not only the plates, the bell-glass and the clockwork need to be carefully coated with vaseline with a brush, but it is necessary to cover the mercury with a thin layer of aseptic glycerine. Moreover the air of the two bell-glasses must be kept constantly saturated with moisture; this may be easily accomplished by placing in a little glass dish under each cylinder a sponge soaked with a saturated sublimate solution.

Hesse's Method for the quantitative estimation of micro-organisms contained in the air is to draw the air through long glass tubes whose walls are coated with stiffened Koch's culture gelatine. The air stream is regulated by means of an aspirator, and at the same time measured. The number of the colonies springing up in the gelatine, and the quantity of air aspirated, give approximate figures for the quantity of germs in the air. Hesse's apparatus consists of a glass tube (as a rule 90 cm. long and 3.5 cm. in diameter), closed at one end by two rubber caps, the inside one having a round hole about 1 cm. in diameter cut in the centre. The other end has a perforated rubber plug 2 cm. thick, through which passes a glass tube with cotton in both ends. The apparatus is sterilized by steam at 100° C. for two hours, and is then coated inside with culture gelatine, which is spread evenly by proper movement, the bulk being allowed to settle on the lower side, the tube being held horizontally. It is allowed to cool, and the tube placed on a tripod such as is used for photography, and the small glass tube passing through the stopper is connected, by means of a rubber tube, with an aspirator made with two one-litre flasks connected by rubber tubes, and hung to the same stand, the upper one filled with water acting as an aspirator. When all is ready, remove the outer rubber cap at the end of the large tube with disinfected hands.

Marchand's Method for the examination of microbes or corpuscles held in suspension in water. The limpid water is placed in a crystal flask covered with black paper, through which are two opposite square openings, one destined for the passage of the luminous fluid, the other for observation. On allowing a ray of light to traverse the water, it does so without obstacles if it is optically pure; but if it has particles in suspension, each one of these is appreciable to sight; but without this device they remain invisible. This method is not new, being that used by Tyndall for testing the optical purity of air.

Certes' Method for ascertaining the presence of micro-organisms at great depths in the ocean. He employed an ingenious device, made by M. Alf. Milne-Edwards, by means of which he could lower small, hermetically sealed, sterilized flasks, and open them at any desired depth, ascertaining at the same time the temperature of the water at that point. The water obtained in this way was added, with all the precautions recommended by M. Pasteur against contamination from atmospheric germs, to culture fluids made of sterilized seawater (120°-128° C.) either alone or in combination with mutton or chicken broth, sterilized hay infusion, milk, or albuminous broths, Raulin and Cohn's fluids, etc. Ærobic microbes were found in abundance, but no anærobic.

LITERATURE OF THE INVESTIGATION OF THE BACTERIA CONTAINED IN THE AIR, WATER, OR EARTH.

BÉCHAMP : " Du rôle et de l'origine de certaines microzymas." Compt. Rend. XCII. (1881), p. 1344–7.

BRAUTLECHT (J.) : (On the transmission of Bacteria from the soil into the air). Tagebl. deutsch. Naturf. Vers. Eisenach. 1882. cf. Naturforscher (1883), p. 156. Journ. Roy. Micr. Soc., Ser. II., vol. III. p. 541.

CASSE (J.) : "Terrain et microbes." 8vo, 18 pp. Bruxelles (A. Manceux), 1884.

CERTES (A.) : I. " Sur la culture à l'abri des germes atmosphérique, des eaux et des sédiments d'apportis par les expeditions du ' Travailleur' et du ' Talisman,' 1882–1883." Compt. Rend. T. 98,

p. 670 (1884). II. "Sur l'analyse micrographique des eaux." Compt. Rend. T. XC. p. 1435. Sep. pub. Paris, 1883. (B. Tignol.) (cf. "Note sur une méthode de conservation des Infusoires." Ibid. séance du 3 Mars, 1879.)

FODOR: "Ueber atmosphærischen Staub, etc." Hygienische Untersuchungen, Braunschweig, 1881.

GIACOSI (P.): (On organic particles in the air of mountains). Atti R. Accad. Sci. Torino. XVIII. (1883), p. 263.

GUNNING: I. "Werden mit der Expirationshift Bakterien aus Körper entführt." Klin. Monats. f. Augenheilk. Bd. XX. Hft. 1. II. "Beitrag zur hygienischen Untersuchung des Wassers." Archiv für Hygiene. Bd. I., 1883, p. 335.

HANSEN (E. C.): (On microorganisms of the air). Resumé du compte rendu des travaux du laboratoire de Carlsberg. 42 fasc. Copenhagne. p. 197–258. cf. Journal de Micrographie, 1882, T. VI., p. 411.

HESSE (W.): "Ueber quantitative Bestimmung der in der Luft enthaltenen Mikroorganismen." Im kaiserlichen Ges., 1884. Bd. 2, Taf. xi.-xiii. II. "Weitere Mittheilungen über Luftfiltration." Deutsche med. Wochenschr. 1884. No. 51.

JORGENSEN (A.) OG HÖYER (H.): "Om Drikkevandet i Kolding" (Microscopical investigation of drinking water). Kopenhagen, 1883. 8vo, 25 pp., with plates and figures in text.

Kidder (J. K.) : "Report on the examination of the external air of Washington." Extr. Report of Surgeon General of the Navy for 1880. Washington, 1882. 24 pp., 10 tables.

Koch (R.) : (Microorganisms in soils) Bied. Centr. 1883, p. 581. 2. cf. Journ. Roy. Mic. Soc. (1884) p. 428. Journ. Chem. Soc. XLVI. (1884) p. 486.

MacDonald (J. D.) : "A guide to the microscopical examination of drinking water. With an appendix on the microscopical examination of the air." 2 ed. (1883). 83 pp., 8vo, 25 tab. Phila. (Blakiston, Son & Co.)

Maddox (R. L.) : "On a portable form of aeroscope and aspirator." Journ. Roy. Mic. Soc., Ser. II., vol. III. P. 2., 1883, p. 308.

Maggi (L.) : Sull' esame microscopico di alcune acque potabile della citta e per la citta di Padova. Pavie, 1883. (Succ. Bizzoni) ("La partie générale constitue un véritable traité d'analyse microscopique des eaux, analyse dans laquelle la récherche des micro-organismes, bactéries et autres, occupe une place importante.") Journal de Micrographie. 1883.

Marchand (E.) : "Sur l'examen des corpuscles tenus en suspension dans l'eau." Compt. Rend. 1883. T. 97, p. 49.

Miquel (P.) : I. "Des bactéries atmosphérique." Compt. Rend. T. 91, 1880. II. "Etude général sur les bactéries d' l'atmosphère." L'Annuaire de Montsouris, 1881. III. "Etude sur les pouss. organisée de l'atmosphère." Nouvelle rect. Betrissonia, III. 5; also, Revue de botanique crypt. III. 2 et 3. IV. "Des procédés employés pour récolter les germes aériens des bactéries." Chapter V. in "Les Organismes vivant de l'atmosphère," Paris, 1883. VI. "Sur le dosage des Bactériens dans les poussières et dans le sol." Bull. de la Soc. Bot. de France. T. XXVIII. (1881). VII. (Hourly Variations of Aerial Bacteria, Miquel's Nutritive Paper). La Semaine Medicale, Nov. 6, 1884.

MIFLET: "Ueber die in der Luft suspendirten Bacterien." Beträge zur Biol. d. Pf., III. 1.

NÆGELI U. BUCHNER: "Der Uebergang von Spaltpilzen in die Luft." Ctbl. f. d. med. Wiss. 1882. No. 29. Absts. in Dtsch. med. Wochenschr. 1882, p. 516.

PASTEUR (L.): "Memoire sur les corpuscles organisés qui existent dans l'atmosphère." Annales de Chem. et de Phys. 1862, Ser. 3, t. 64. Also Compt. Rend. T. 48, 1859, p. 337; T. 50, 1860, p. 849.

PASTEUR ET JOUBERT: "Sur les germes des bactéries en suspension dans l'atmosphère et dans les eaux." Compt. Rend. T. 84, p. 206 (1877).

PÖHL (A.): "Chemische und Bakteriologische Untersuchungen, betreffend die Wasserversorgung

St. Petersburgs nebst einem Beitrag zur Biologie der Microorganismen." Petersburger med. Wochenschr. 1884, No. 31 u. 33. — Journ. Russian Chem. Soc. 1884.—Nature, XXIX. (1884), p. 557.

SCHRÖDER U VON DUSCH: "Ueber Filtration der Luft in Beziehung auf Faulniss und Gahrung. Annalen der Chemie und Pharmacie." 1854. Bd. 89, p. 232.

SCHUTZ: "Ueber das Eindringe von Pilzsporen in die Athemungswege und die dadurch bedingten Erkrankungen der Lungen und über den Pilz des Hühnergrindes." Mitt. a. d. kais. Gesundheitsamt. 1884. Bd. 2.

TIEMANN: "Untersuchung des Wassers auf entwicklungsfähige Mikroorganismen." Verhandl. dtsch. Gesellsch. f. offentl., Gesundheitspflege zu Berlin, 1883.

TISSANDIER (G.): "Les poussieres de l'air." Paris (Gauthier et Villars), 1877. 1 vol., 18mo.

TYNDALL (J.): "Essays on the Floating Matter of the air." 2d edit. 1883.

WERNICH (A.): I. "Ueber das Haften und die Ansiedlungsfähigkeit staubförmige Pilzkieme." Dtsch. med. Wochenschr, 1882, p. 513. II. "Die Luft als Trägerin entwicklungsfähiger Keime." Virchow's Archiv. Bd. 79, p. 424–445.

WOLLNY (E.): "Ueber die Tatigkeit niedere Organismen im Boden" Deutsche Vierteljahrsschr. f. œffentl. Gesundheitspflege, XV. 4.

If a person desires to isolate for investigation, from some fluid containing several forms of Bacteria, some one particular form, he may best proceed according to —

Nægeli's Dilution Method, as illustrated in the following example. Prof. Nægeli had some urine containing large *Cocci* and also numerous *Bacilli*; the former he desired to obtain pure. A drop of the urine (about 0.03 ccm.), estimated to contain 500,000 Bacteria, was placed in 30 ccm. of pure sterilized water; after being thoroughly shaken, one drop of this thousand times diluted urine was added to a second 30 ccm. of water, and thus a millionth dilution was obtained, in which every two drops of 0.03 ccm. must contain one microbe. Each of ten tubes of sterilized culture fluid [*vid.* CULTURE METHODS] was then inoculated with one drop of this diluted urine, and after proper incubation it was found that four tubes remained sterile, one contained *Bacilli*, and five the desired *Cocci*. These he could now study alive as microscopical preparations, or by any method open for the investigation of Bacteria. — On the other hand, the Bacteria which it is desired to study may not be in a condition suited for microscopic study, they may be dried or in the "resting spore" condition, requiring special treatment before they can be studied alive, undergoing the various phenomena of their life history; thus —

Buchner's and Robert's Method for Bacillus subtilis (the Hay Bacterium) is to digest a quantity of finely-chopped hay for an hour, with as little water as possible, at a temperature of 36° C. The liquid is then drained off through a wire sieve, and diluted with distilled water until it reaches a specific gravity of 1.004. If the fluid is acid in reaction, it must now be neutralized with carbonate of soda, and not less than 500 ccm. placed in a flask, which is stopped with a plug of cotton. The flask and contents are now retained for forty-eight hours at a temperature of 36° C., when the developed *Bacilli* will have formed a dry-looking pellicle upon the surface.

BACTERIA IN HEALTHY, LIVING ORGANISMS.

Peculiar procedures had to be employed for developing the bacterial germs which exist in the tissues of healthy living animals, and rendering them suitable objects for microscopic examination. The first of these was —

Tiegel's Method. The organ to be examined, or a piece of it, is removed with a sterilized knife from an animal just killed (usually by bleeding through the carotid). A silk thread, previously well boiled, is quickly tied about the piece, and by means of this it is dipped in melted paraffine at 110° : 115° C., and allowed to remain a longer or shorter time, according to the size of the piece. After the paraffine remaining on the surface has cooled, it is again dipped and very quickly withdrawn. This is only intended for the purpose of strengthening the paraffine crust. After this layer has cooled, the preparation is dropped into a mass of cooling paraffine at 52° C., and here allowed to cool. The lump thus formed is then for a length of time, 4 to 12 days, retained at a temperature of about 30° C., then cut open and the interior studied. The strong scalding which the organ receives in the first paraffine kills such microbes as may have fallen upon it during its removal, and before they can have reached the interior. Billroth and Tiegel always found Bacteria in the pancreas, liver, spleen, salivary glands, testicles, and muscles after 4 to 12 days; most in pancreas.

Burdon Sanderson's Method, to overcome the objection that the Bacteria might have gained entrance to the tissue through the cracks and pores formed in the paraffine while cooling, was to throw the organ into melted paraffine at 110° C., and, as soon as it was cool and stiff, to paint it over with Venetian turpentine. Sometimes he simply dropped the organ into oil at 110° C. and allowed it to remain at the bottom of the vessel for 1 to 2 days. The surface was found to be cooked, but the central portion retained the pale red color of raw tissue, and contained numerous Bacteria at different stages of development. Richtet employed this method in seeking Bacteria in fish. Thus, in the case of a live eel, he opened it with sterilized scissors, cut out its liver, and plunged this into melted paraffine at 110° C. After cooling, the mass was covered with several coats of collodion, and then with Canada balsam. Three weeks after, the liver contained myriads of small *Bacilli*, but it still had the odor of fresh fish.

Chiene and Ewart's Method: Under a spray of 5 per cent carbolic acid solution, the body of a rabbit, just killed, was opened, and the liver, spleen, kidneys, and pancreas taken out and cut into pieces. Some of these were dipped into a solution of carbolic acid, others were wrapped in antiseptic gauze; they were then placed in vessels which had been heated red hot, and the openings closed with antiseptic gauze, cotton, or with a glass cover. In those pieces which had been dipped into the antiseptic solution they found no Bacteria, and hence concluded that there were none present in the organs. This gave rise to —

Nencke and Giacosi's Methods, proving that Chiene and Ewarts had killed the Bacteria with carbolic acid.

(*a*) In a beaker-glass of the capacity of about one-half litre, a quantity of some alloy which is readily melted, say at 75° C. [Wood's metal], is heated up to 300° or 400° C. As soon as the alloy has cooled to 150° C., a layer of 5 per cent carbolic acid solution is poured upon it, and the metal is retained in a melted condition by being placed in a water-bath. Now, a rabbit is killed, and its abdomen shaved and washed with the carbolic acid solution. The abdominal cavity is then opened under a spray of 5 per cent carbolic acid solution, and a large piece of the liver cut out and placed with sterilized pincers in the melted Wood's metal, and the whole allowed to cool and stiffen about both tissue and pincers. The mass is then kept for four days at 40° C. In this method there is no danger of the material enclosing the tissue cracking, and, if this should happen, no germs from the air could reach the enclosed tissue on account of the intervening 5 per cent carbolic acid solution.

(b) *Nencki and Giacosi's* second method is to take a large enamelled iron vessel, and fill it one-third full of pure mercury. A glass tube about 5 ctms. in diameter is taken, and one end melted round, like the fundus of a test tube, and the open end ground perfectly even. This is then filled with mercury, closed with a glass plate, and inverted into the vessel, being allowed to remain leaning against the side. The whole apparatus is then heated until the dish is about two-thirds full of mercury. Any germs which might be in the tube are killed by the boiling mercury. It is now allowed to cool, whereby the mercury is again condensed in the tube, and, when the temperature reaches the neighborhood of 120°, a 5 per cent solution of carbolic acid is poured over the surface. With proper precautions and the use of the carbolized spray, some organ of a freshly killed rabbit (liver, heart, kidney, spleen, pancreas) is taken and placed with the pincers under the mouth of the tube, to the top of which it rises, and there remains. The preparation is now kept at a temperature of 40° C. for one or more days. In these methods all organs investigated showed the presence of Bacteria in large numbers.

LITERATURE ON THE PRESENCE OF BACTERIA IN HEALTHY LIVING ANIMALS.

BABES (V.) : (No Bacteria in blood or tissues of healthy men). Biol. Centralbl., 1882, II., p. 97–101.

BALZER (Leptothrix epidermadis): Annales de Dermat et de Syphilographie, 25 Dec., 1883, p. 681.

BÉCHAMPS (J.) : I. "Les Microzymas" (Micrococci). Montpellier et Paris, 1875. II. "Des microzymas gastriques, et de leur pouvoir digestif." Compt. Rend. 27 Feb., 1882. (Journal de Micrographie, T. IV., p. 188, 1882.

BÉCHAMPS (A.), ESTOR (A.), ET SAINTPIERRE (C.) : " Du rôle des organismes microscopiques de la bouche dans la digestion en général, et particulièrement dans la formation de la diastase." Compt. Rend. T. LXIV., p. 696 (1867).

BEIMSTOCK (B.): "Ueber die Bakterien der Fæces," Fortschritte der Med., 1883, No. 19; also Zeitschr. f. klin. Med. VIII.

BILLROTH und TIEGEL: "Ueber Cocco Bacteria septica (Billroth) im gesunden Wirbelthierk rper." Virchow's Archiv. f. pathol. Anat. u. s. w. Vol. 60, p. 453.

BIZZOZERO (G.): "Sui Microfiti dell' Epidermide umana normale," Estratto dal volume d'Atti. della R. Accad. di medicina di Torrino, 1884. Cf. Virchow's Archiv., 1884, Bd. 98, p. 441.

BURDON-SANDERSON: "Bacteria in healthy living organs." British Med. Journ., Jan. 26, 1878.

CARL (Herzog v. Bayern): "Bei Menschen vorkommende Bacillen" (in the choroid). Abth. f. med. Wiss., 1881, I. Louisville Med. Herald, iii. (1881), p. 27, Journ. R. Mic. Soc., Ser. II. vol. vi., p. 644.

CHIENE (J.) and EWART (C): "Do bacteria or their germs exist in the organs of healthy living animals." Journ. of Anat. and Physiol., vol. xiii. part 3, p. 448, 1878.

DOWDESWELL (G. F.): "On some appearances in the blood of vertebrated animals, with reference to the occurrence of Bacteria therein." Jour. Roy. Micr. Soc., 1884, p. 525–529.

ESTOR (A.): "Contribution à l'étude des microzymas et des bactéries." 8vo, 20 pp. Paris (Delahaye et Lecrossier), 1884.

FRISCH (A.): "Experimentelle Studien üb. d. Verbreitg. d. Faulnissorganismen in d. Geweben." mit. 5 Tafln, 4to, 1874.

HOFFMAN (G. v.): " Untersuchungen über Spaltpilze im menschlichen Blute, Ein Beitrag zur allgemeinen Pathologie." Gr. 8, mit 2 lith. Tafl. 1884.

KÜNSTLER (J.): (*Bacterioidomonas sporifera* in cæcum of *Cavia*). Journal de Micrographie, 1884, VIII., pp. 376-80. Jour. Roy. Mic. Soc., 1884, p. 934.

LEUBE (W. O.): "Beitrag zur Frage von Vorkommen der Bakterien in lebender Organismus, speciell im frischgelassenen Harn." Zeitschr. f. klin. Med. III., p. 233.

LEWIS (T. R.): I. "Microscopic Organisms found in the Blood of Man and Animals." Fourteenth Annual Report of the Sanitary Commissions with the Government of India. II. "Microphytes in the Blood, and their Relation to Disease." Quart. Journ. Mic. Sci., XIX. (1879), p. 356.—Journ. Roy. Mic. Soc., 1879, p. 924.

LUDWIG (F.): I. (*Micrococcus prodigiosus* within the shell of an egg.) Zeitschrift f. Pilzfreunde, 1883, p. 176.—Botan. Ctbl. lxviii. (1884), p. 160. Journ. Roy. Mic. Soc. (1884), p. 596. II. (Photogenous Micrococci, *Micrococci pflugeri* the cause of phosphorescence in fish, Crustacea, echinoderms, etc., and of phosphorescent patches on the surface of the sea.) Hedwigia, XXII. (1884), p. 106–15. Journ. Roy. Mic. Soc., 1884, p. 596.

METSCHNIKOFF (E.): "Untersuchungen über die Intracellulare Verdauung bei Wirbellosen Thieren" (Bacteria in Tunicates fresh from the sea, p. 21), Wien, 1883. (Transl. in Quart. Journ. Micros. Sci. for Jan., 1884.) II. "Ueber die pathologische Bedeutung der intracellulären Verdauung," Virchow's Archiv, Bd. XCVI., Hft. 2, 1884. cf. Arbeiten des zoolog. Instit. zu Wien, Bd. V., Hft. 2, 1883. — Biologisches Centralbl. Bd. III. No. 18. 1883.

MEYER (L.): (On rod-like bacteria found on the mucous surfaces of the female sexual organs). Vortrag in der Ges. fur Geburtshilfe, Berlin, 1862.

MILLER (W. D.): "Zur Kenntniss der Bakterien in der Mundhöhle," in Deutsche med. Wochenschr. No. 48, 1884. p. 781.

Moss (E. L.): "On septic organisms in living tissues." Journ. Roy. Mic. Soc. (1879), p. 312.

NENCKI (M.) UND GIACOSA (P.): "Giebt es Bacterien und deren Keime in den Organer gesunder lebener Thiere," in Beiträge zur Biologie der Spaltpilze, Leipzig, 1880.

NOTHNAGEL (H.): "Bacillus Amylobacter (Clostridium butyricum) im Darwinhalt." Ctbl. f. d. med. Wiss., 1881, p. 19.

PETER. (Bacteria as the cause of the various coloring of the eggs of *Coregonus wartmanni*.) Berl. bot. Verein, München, Sept. 19, 1883. Journ. Roy. Micr. Soc., 1884, p. 601.

RASMUSSEN: "Om Drychning af Mikroorganismer fra Spyt af Sunde Mennesker" (On the culture of the micro-organisms of the saliva of healthy men). Kopenhagen, 1883. Dissertation.

REYMOND: "Batteri del sebo delle ghiandole meibomiane normale." Giornale dell' Acc. di Med. di Torino. (Lugli.), 1883.

RICHET (Ch.): "Les microbes des poissons marins." Comptes Rendus de la Acad. des Sciences de Paris. Tome 96, p. 384 (1883).

ROBERTS (W.): "On the occurrence of microorganisms in fresh urine." Brit. Med. Journ., 1881, Oct. 15.

ROCHAS (F.): "Les schizophytes parasite de l'homme et des animaux." Lyon. (Georg), 1885, pp. 27, 8vo.

ROSBACH (M. J.): "Ueber die Vermehrung der Bakterien in lebenden Thiere nach einverleibung eines chemischen organismen freien Fermentes." Ctbl. f. d. med. Wiss., 1882, No. 5.

STERNBERG (G. M.): "A contribution to the study of the bacterial organisms commonly found upon exposed mucous surfaces and in the alimentary canal of healthy individuals." Studies from the Biol. Lab. Johns Hopkins University. II. (1882), pp. 157–181.

TRINCHESE: (Bacteria in the Human Amnion.) Atti R. Acad. Lincei. 1884, p. 237. Journ. Roy. Micros. Soc. (1884), p. 268.

VULPIAN: Bull. de l'Acad., 29 Mars, 1881.

WASSILIEF (N. P.): "Beitrag zur Frage über die Bedingungen unter denen es zur Entwicklung von Mikrokokken. Colonien in den Blutgefassen kommt." Ctbl. f. d. med. Wiss. 1881. No. 52.

ZAHN: "Untersuchungen über das Vorkommen von Faulnisskeimen im Blut gesunder Thiere." Archiv. f. pathol. Anat. 1884. Bd. CXXXV., p. 401.

To obtain blood for microscopical examination, remove the hair from the spot to be punctured, clean the skin carefully with warm water and soap, place a drop of 0.8 per cent salt solution on the spot, and with a knife, which has been previously heated, cut through the skin; then with another knife cut a small vein or artery.

Moss's Method for obtaining blood for bacteriological investigation. This method seems to exclude all possibility of infection, and at the same time allows the blood to be examined at intervals. The apparatus consists of a series of small glass bulbs connected together by capillary tubes, so that one bulb and its contents can be separated from the rest by the blow pipe. The tubes and bulbs are bent on each other, so that the whole series can be readily sterilized in a water or paraffin bath. One end of the series is left open, packed with sterilized cotton, and connected with an aspirator, the other drawn to a fine point and sealed. The sealed point is secured in a stout piece of india-rubber connection pipe, which is fastened over the collar of a fine hypodermic needle, protected ready for use in a calcined glass sheath. The whole arrangement is then repeatedly baked in a water bath at intervals of four hours. (Tyndal's method of sterilization by discontinuous heating.) In using the apparatus the sheath is removed from the needle, and the latter plunged into any suitable vein. The sealed point inside the rubber connection-tube is broken, and the blood flows gently through the series of bulbs, drawn on by the aspirator acting through the cotton plug. When sufficient has entered, the blow-pipe severs the tube next the needle, and instantly afterwards that next the cotton plug.

Only in two diseases are Bacteria regularly to be found in the blood of living persons, *i. e.*, in *Anthrax* and *Febris recurrens*. The anthrax *Bacilli* withstand the action of most reagents, while the *Spirochœte* of recurrent fever are readily destroyed by the addition of fluids, even by distilled water.

To obtain serum or fluid from a freshly extracted

organ, cut out a piece with a previously heated knife, and with another sterilized knife cut through this piece, and press a *clean* cover-glass against the freshly cut surface; or by scraping the surface with the sterilized platinum wire, sufficient fluid may be obtained, and transferred to the cover-glass. If the cut surface of the organ or tissue be of a consistency not easily spread out, a few drops of sterilized water should be placed upon the freshly cut surface before laying the cover-glass upon it or scraping it with the wire.

The Study of Microscopical Preparations of Fixed and Hardened Bacteria, while not by any means the most important, is the method which has been most largely employed. When the Bacteria are to be examined in opaque substances, such as fresh tissues, mucus, or fluids containing much detritus, they must be rendered visible upon the slide by means of clearing agents. Fresh tissues (tissues which have been for a length of time in alcohol or other hardening fluids are not readily cleared up, and therefore not easily examined without staining) should be examined as early as possible after removal, in sections cut with a freezing microtome or a Valentin's knife; these, as well as turbid fluids, should be spread out upon the slide, and cleared up with either a 15 per cent sulphuric acid solution, a 20 per cent solution of caustic potash, concentrated acetic acid, absolute alcohol, or ether, as may seem proper for the case in hand. Usually the microbes withstand the action of the clearing fluids, while the tissue or detritus almost entirely disappears. Examine these preparations with a narrow light. It is not infrequently the case that sections of tissues from cases of metastatic pyæmia, endocarditis ulcerosa, and acute croupous pneumonia show the blood-vessels, capillaries, small veins, and, in the last-named disease, the lymph vessels, to be irregularly filled with heaps or chains of a granular material. These "capillary emboli" may consist of *Micrococci* colonies, which may be considered proven if we find that they withstand the action of clearing agents, as indicated above. Control tests should be made by staining some sections.

The Use of Staining Fluids, in connection with Bacteria which have not been hardened or fixed, is not satisfactory; and yet they are often still less easily stained after having been for a long time exposed to the action of alcohol or other hardening agents, or after having been for a long time dried. The best results are to be obtained from tissues which have remained for only a short time in hardening fluids. Staining fluids are employed either for the so-called "cover-glass preparations," in which the Bacteria to be studied exist in fluids, or for the preparation of sections in cases where the substratum is a firm one. Bacteria behave in staining much like nuclei, and, as a rule, such staining fluids as are good for nuclei will succeed with Bacteria. On account of their minute size, however, the more intense the stain, the better; therefore the aniline dyes seem best suited. The intensity of the staining can be increased by keeping the preparations at a temperature of about 50° C., in a drying-oven. The behavior of Bacteria and nuclei is very different towards alkalies; the nuclei are dissolved by alkalies, while Bacteria are unaffected; therefore the staining fluids are often made alkaline in order that any confusion of Bacteria with nuclei may be avoided. For *Micrococci* most nuclear stainings may be used. Koch's and Gram's general methods, or Lœffler's method given under diphtheria, give good results. It is known, however, that living *Micrococci* take staining much better than do dead ones. It is possible that further study of the way in which different species of *Micrococci* take staining may show characteristic specific peculiarities, but as yet there are none known; the *Micrococci* of malignant endocarditis, of pyæmia, of erysipelas, and of gonorrhœa, etc., behave alike. Weigert recommends, *for red*, all carmines (Schweigger-Seidel's), purpurin, fuchsin, or Magdala-red, *for brown*, Bismarck-brown or vesuvin; *for brown-violet*, carmine, followed by washing in alcohol to which has been added a little liquor ferri sesquichloride; *for green*, methyl-green; *for blue and violet*, hæmatoxylin, iodine-violet, methyl-violet, gentian-violet, or dahlia. He recommends for *Bacilli* the nuclear staining aniline dyes, those, for instance, which Ehrlich designates "basic anilines,"— Bismarck-brown, methyl-violet, methyl-green, saffranin, fuchsin, Magdala-red, etc. Excelling all these, however, for *Bacilli*, is gentian-violet. Use all the anilines in strong aqueous solutions until the preparation is over-stained, and remove the excess by washing in acetic acid, water, or in alcohol. Carmine and hæmatoxylin are not recommended for *Bacilli*, although Klebs used the latter with good success in his researches regarding the *Bacillus* of typhoid fever.

Hazlewood recommends, as a general blue stain, a mixture of rosanilin, aniline oil, and sulphuric acid, as giving "surprisingly fine results with *Micrococci*, *Bacilli*, etc. Whether the mount is in balsam or glycerine, the result is excellent, though the glycerine mounts are in some respects preferable."

Blanchard's Method: In a few hours, or two days at the longest, the surface of water in which an organized substance (vegetable or animal tissue, etc.) has been macerated, becomes, as is well known, covered with a slight pellicle, composed of a more or less compact mass of Bacteria, enveloped in a hyaline, transparent substance of slight consistence. A tolerable large piece of this membrane ("proliferous membrane" Pouchet) can be obtained by introducing into the fluid beneath it a glass slide, and raising it with caution. The film thus obtained is fixed by treating it with strong osmic acid. It is then covered, the osmic acid drawn off, and a drop of methyl-violet run in under the cover. In half an hour's time the preparation may be completed by running in glycerine, to which is added a small quantity of the violet, in order that the stain may not be extracted from the organisms; or a concentrated solution of sulphate of calcium may be used instead of glycerine. The Bacteria are stained a fine violet, the ground substance remaining colorless. Other aniline dyes may be used, but methyl-violet seems to be the most durable. Hæmatoxylin may also be used. In this case, the film of Bacteria should be stained in it for twenty-four hours (after fixing with osmic acid); the iridescence which is then formed, and which spoils the clearness of the preparation, is removed by repeated washing, and the membrane is mounted in glycerine, or chloride of calcium.

"*T. C.*"[1] recommends the following method of obtaining permanent preparations of Bacteria, it being the result "of some years of patient research." A ring of white wax much larger than

---

[1] Science Gossip, 1879. No. 173, p. 111.

the cover-glass is drawn upon a slide, within which the solution containing the microbes is placed. When they have attached themselves to the slide, some of the fixing solution (25 cc. of chromic oxi-chloride acid, to which is added 50 cc. of water with 50 cc. of permanganate of potash) is added, which will instantly kill and fix the specimens; after three minutes, the water may be poured out, a few drops of absolute alcohol and then of chloroform added and poured off, the cover-glass placed carefully on, and some thin Canada balsam run under.

It is important, in the use of aniline dyes, that the materials used should be of a good quality, and, if it is desired to compare results with those obtained by European investigators, it is well to use those dyes prepared especially for Bacteria investigation.[1] A small piece of camphor placed in the aqueous solutions of aniline dyes does not impair their staining qualities, and tends to prevent the growth of moulds. These solutions must, however, always be filtered before using.

Cover-glass Preparations, for the study of Bacteria contained in fluids, by fixing, hardening, and

---

[1] A collection of the more important aniline dyes can be obtained of Dr. Georg Grübler, "Physiologisch-Chemisches Laboratorum," Leipsic, Durfour Strasse, No. 17, for about $4.00, by ordering through any of the New York, Philadelphia, or Boston drug houses, or by sending a money-order direct to Dr. Grübler.

staining them. Always make more than one preparation, usually three. Place a small quantity of the fluid to be examined upon a sterilized cover-glass (*i. e.*, one which has been passed two or three times through a spirit flame), cover it with another and *slide* the two apart (do not lift them apart). Then place them in a drying-oven, at a temperature of 120° C. Simple exposure to the air will answer in some cases. By far the best method, however, since the Bacteria and cellular elements contained in a thin layer of fluid are not injured by rapid drying, is to seize the cover-glass upon which is spread the fluid to be examined, with the pincers, and to pass it three or four times with the prepared side up, to and fro through the flame of a Bunsen burner or of a spirit-lamp, "about as fast as you would cut bread" (Rindfleish). After a few trials one learns how to do this without burning the preparation. This heating coagulates any albumen present, and prevents the specimen swelling up in the subsequent treatment. When it is desired to retain all the finer details of structure, which are usually lost by drying, specimens may be fixed and hardened by exposing them to the fumes of osmic acid, or, still better, to the fumes of absolute alcohol. Pour a few drops of absolute alcohol into a deep watch-glass; suspend the cover-glass over this, with the prepared side up; place a second watch-glass over the whole, and take it all up with the pincers and warm it over the flame until it steams slightly. Besides the above, the ordinary methods of hardening may be employed, — chromic or picric acids, or their combinations with other acids. The prepared cover-glasses should be thoroughly rinsed of these acids before staining.

General Methods for Staining Cover-Glass Preparations. — In those cases when it is not desired to show the characteristic manner in which certain pathogenic Bacteria behave towards staining fluids, the preparations may be stained with Bœhmer's hæmatoxylin [F. 22.] with methyl-blue or with a concentrated aqueous solution of vesuvin [F. 38.]. Where a more intense color is desired, or in working with those forms which do not stain readily methyl-blue [F. 23.], gentian-methyl-violet [F. 18.], or acid fuchsin [F. 37.] may be used in connection with a watery solution of aniline oil, [F. 5.]. In any of the above (*Koch's Method*) allow the cover-glass to float prepared side down upon the surface of a quantity of the staining fluid in a watch-glass, for from two minutes to twenty-four hours (raising the temperature of the staining fluid shortens the time required); remove, and wash in acidulated water [F. 7.] until the color has apparently faded out, pass rapidly through absolute alcohol, dry, and mount in Canada balsam (free from chloroform) or acetate of potash solution [F. 34.].

Of all general methods for staining Bacteria in cover-glass preparations, Friedländer considers that of Dr. Gram, of Copenhagen, to be by far the best and nearest perfection. It is especially useful where the material to be investigated is rich in nuclei; as blood, pus, and the splenic or hepatic juices. It allows of an isolated staining of the Bacteria, everything else in the field remaining colorless; the Bacteria being so intensely colored a deep dark blue that the investigator must find every single individual present.

Gram's Method. — The prepared cover-glasses are allowed to float prepared side down upon a quantity of filtered Balmer-Fränzel's aniline-gentian-violet solution [F. 19 a.] which is in the meantime held over a flame and warmed. (The preparations used in this method must have previously been in absolute alcohol, not in water or dilute alcohol.) They are then placed for one minute in iodo-iodide of potash solution [F. 35 d.], then washed in absolute alcohol (1–3 min.) until the color becomes invisible to the naked eye, clear up in oil of cloves, and mount in xylol-Canada balsam or in glycerine-jelly [F. 20.]. Special methods for staining "cover-glass" preparations will be found in their appropriate place under the treatment of pathogenic Bacteria. Beautiful double-stainings may be produced by the use of Bismarck-brown after Gram's method, *i. e.*, after being stained violet and washed in the potash solution and alcohol, place the cover-glasses for a moment in Bismarck-brown or vesuvin solution, then wash in absolute alcohol, pass through oil of cloves, and mount. The Bacteria remain a dark blue, while the nuclei become a yellowish brown.

The Preparation of Bacteria in and upon Firm Substances. — If the tissue to be examined is a fresh one, cut sections of it with a freezing microtome or a Valentin's knife, and pass them, before staining, through absolute alcohol. Hardened tissues may be cut in serial sections by the microtome, and left in alcohol until desired for staining. Only small pieces of tissue should be taken for hardening, and should be placed in relatively large quantities of alcohol. Hardening in chromic acid or Müller's fluid is not well suited for the study of Bacteria, since these fluids are liable to produce in the tissues numerous dark granules which are difficult to clear up. The hardening must not be continued too long, as many Bacteria lose their capacity for taking stain after being three or four days in alcohol. Neither must the specimens to be cut, be imbedded in paraffine, as this and the volatile oil requisite to the imbedding process destroys the staining capacity of the microbes. Transparent soap [F. 41.] or celloidin [F. 42.] may be used as imbedding masses. Two general plans are open for staining the Bacteria which exist in sections of firm substances, *i.e.*, that in which but one color is used, and the tissue substance afterwards bleached out, leaving the stained Bacteria upon a transparent ground, or that in which a double staining is employed, the microbes retaining one color and the tissue another. After the first plan we have —

Weigert's Method, according to which a solution of any basic aniline dye may be used, methyl-blue [F. 23.], methyl-violet [F. 25.], gentian-violet [F. 18.], rosanilin-chlorhydrate [F. 36.], magenta-red [F. 26.], fuchsin [F. 16.], Bismarck-brown [F. 2.], vesuvin [F. 38.] and aniline-brown [F. 3.] being the better, while iodine-green, malachit-green [F. 24.], and chrysoidin [F. 11.], aniline-black [F. 27.], and aniline-blue are less preferred. Allow the sections to remain for 18 to 24 hours in a staining fluid prepared from one of the above; the process may be hastened by warming the solution at 45° C.; it is, however, easy to over-stain by doing this, although such an accident can be rectified by subsequent treatment with a saturated solution of carbonate of potash. After staining, wash in distilled water, then place for 5 minutes in 50 per cent alcohol, and then for 15 minutes in absolute alcohol, in order that all water may be removed, and the tissue decolorized. Finally clear up in oil of cloves, oil of bergamot, or xylol, and mount in xylol-Canada balsam. Another plan adapted to all forms of Bacteria is —

Schütz' Method, which can be used in sections of fresh tissues cut with the freezing microtome. Place the sections for twenty-four hours in a methyl-blue solution [F. 23.], wash in water to which a few drops of acetic acid have been added, then in 50 per cent alcohol for 5 minutes, absolute alcohol 15 minutes, clear up in cedar oil, mount in Canada balsam.

Babes recommends the use of safranin for staining sections of pathological specimens containing Bacteria. The sections, hardened in alcohol or chromic acid, should be steeped in a super-saturated solution of saffranin, which is warmed to 60° C. and filtered in this state. The sections are placed in a small quantity of the liquid in a watch-glass, which is then warmed for a few seconds over a spirit lamp until the precipitating crystals are re-dissolved; the sections are left for a minute, then washed in water, and dehydrated in absolute alcohol as quickly as possible, then transferred to turpentine, and mounted in balsam. Every *Micrococcus* appears brownish-red, while the surrounding tissue assumes a fine rose-red; the *Bacilli* of tuberculosis and lepra are not thus stained.

According to the second general plan, that of *Double Staining*, we may, after removing from the tissue the color first employed, pass the sections through a solution of some other aniline dye: the browns are to be preferred, *e.g.*, Koch's "glycerine-aniline-brown" [F. 3.]. It is not absolutely necessary to remove the first color from the substance of the tissue, as this will retain the color to which it is last subjected; the sections may therefore be passed at once from the first staining fluid into a second filtered aqueous solution of almost any of the basic aniline dyes, which will change the color of the tissue, while the Bacteria retain the color to which they were first subjected. Of course only complementary colors should be chosen for double staining. The following make good combinations, methyl-blue and vesuvin [F. 23, 38.] (used by Koch in cover-glass preparations), gentian-violet and vesuvin [F. 38.], or picrocarmine-fuchsin and methyl-blue [F. 23.] or *Subbotin's Combination*, in which he took preparations which had been hardened by means of osmic acid fumes [F. 30] or with chromic acid, and, after washing, treated them with a one per cent aqueous solution of methyl-green [F. 24.] for about two hours, and subsequently with picro-carminate of ammonia [F. 33.], washed them again, dried and mounted them in Canada balsam. The Bacteria appeared green, while the tissue was stained red.

Mounting. — Such specimens as are desired for

purposes of photo-micrography, if colored in vesuvin (best adapted), should be mounted in glycerine, or glycerine-jelly may be used [F. 20.], which becomes fluid by slight warming, and stiffens again when cool; if not colored with vesuvin, the saturated solution of acetate of potash [F. 34.] should be used for mounting specimens to be photographed. The cover-glasses can be immediately secured in place in glycerine mounts by touching a heated wire to a piece of paraffine or wax, some of which will form a drop at the end of the hot wire; allow this to fall at one corner of the cover-glass, thus you fix the latter so that it is not readily moved while more paraffine or wax is applied, and spread over the edges with the hot wire, or, better than paraffine or wax is the cement recommended by Prof. Csokor [F. 12.]. These are all preferable to "gold size," since by their use the cover-glass can be washed and wiped in a few minutes after mounting without fear of displacing it. In place of these substances a good thick solution of shellac in alcohol, or of Canada balsam in chloroform, or asphalt varnish may be used. Aniline colors, other than vesuvin, do not keep well when mounted in glycerine; Canada balsam is therefore used in most preparations, which, if they have been treated with aqueous staining fluids, must be dehydrated by drying, or by passing through absolute alcohol, then cleared up with oil of cloves, oil of bergamot, or xylol [F. 29.], and finally mounted.

PREPARATION OF BACTERIA FOR PHOTOGRAPHING.

Sternberg's Method of preparing Bacteria for photographing. The Bacteria are dried upon a slide or a thin glass cover, and are then treated with commercial sulphuric acid, a drop of which is placed upon them. After two or three minutes the acid is washed off by a gentle stream of water, and the Bacteria are then covered with an aqueous solution of iodine (iodine, grs. 3, potassic iodide, grs. 5, water, grs. 500). After a few minutes they will be found to present a deep orange or brown color, which gives the desired contrast in a photograph negative. This method is only useful for extemporaneous preparations which are to be photographed immediately. The color fades after a time, and the Bacteria undergo changes in form (swelling) as a result of this treatment, which renders the method unsatisfactory when the object is to make a permanent preparation. For this purpose nothing is better than aniline-violet, which, indeed, leaves nothing to be desired when a collection is being made without reference to photography. The specimens should be mounted either in a solution of acetate of potash (Koch's method), or preferably in carbolic acid.

Kaschka's Method of preparing Bacteria for

photographing. After the drop containing the Bacteria is dried upon the slide in the usual manner, the spot is moistened with a metallic solution of an iodide (cadmiumiodide 1 : 50 was employed), and in two or three minutes the Bacteria are sufficiently iodized. The slide is then carefully and rapidly washed in distilled water, and immediately flowed with a few drops of silver solution from the negative bath. If the right time has been hit, and the iodide has not acted for too short a period, and the washing has not been continued too long, the contour of the dried drop will be seen to show a slight yellow color, due to the iodide of silver which is formed. Only an exceptionally short exposure to light is sufficient, after which the developer (strongly acidified and dilute iron developer) is added, and the drop suddenly becomes black. After thorough washing, the deeply-colored Bacteria are mounted in balsam, and may then be readily photographed. This method of staining is only useful for photographic purposes, and there is some chance of mistaking fine silver precipitates for *Micrococci* or other forms. In case of any doubt of this kind, the original forms should be stained with aniline colors and examined in the usual way.

Dufrenne's Method of photographing Bacteria stained with fuchsin. M. Dufrenne describes the process which he adopts for photographing Bacteria with a Tolles 1-10th inch (homogeneous immersion) objective, without eye-piece, and the use of extra rapid bromo-gelatine plates, developed with ferro-oxalate; a petroleum lamp being employed for illumination. "If," he says, "the determination of the actinic focus of objectives constitutes, so to say, the chief difficulty in photographing microscopic preparations, it is no longer so when we deal with organisms so infinitely strong as *Bacilli*. Here arises a difficulty of quite another kind, which at first seemed insurmountable — the staining the *Bacilli* by means of fuchsin. This agent, even when it is employed in thick layers, is somewhat actinic, and it becomes more so as the object stained is smaller or more transparent. These two circumstances are combined in the highest degree in the organisms in question. Thus, at the beginning, the plates exposed were either uniformly acted on, or the image was so faint and so little differentiated after development, that they were worthless for proofs on glass or on paper. These negative results suggested the abandonment of the attempt, when the idea was suggested of having recourse to the use of a *compensating glass* of a color complementary to red (that is, green), placed between the objective and the sensitized plate. By thus filtering the image formed by the objective, the red rays — the only ones passing through the *Bacilli* — are absorbed, if not wholly, at least in great part. The microbes, therefore, appear nearly black on the plate, and make a much slower impression than the rest of the preparation which gives free passage to all the green rays. More contrast is thus obtained, and a very distinct photograph produced."

LITERATURE ON THE PREPARATION AND STUDY OF BACTERIA IN GENERAL.

ABBÉ: "Ueber Blutkorperzahlung," Sitzungsbericht der Jenaischen Gesell. f. Med. u. Naturwissenschaft. 1878. No. 29.

ADAMS (J. M.): "Easy Method of Staining Bacteria." The Microscope, IV. (1884), pp. 224, 225.

BABES (V.): "Ueber einige Färbungsmethoden, besonders für krankhafte Gewebe, mittlst, Safranin und deren Resultate." Arch. f. mikr. Anat. XXII. (1883), pp. 356–65.

BIENSTOCK: (On staining). Zeitschrift. f. klin. Med. 1884, p. 1.

BLANCHARD: (On method of staining). Rev. Inter. Sci., III., 1879, p. 245. — Journ. Roy. Mic. Soc., II., 1879, p. 463.

BUCHNER: "Ueber das Verhalten der Spaltpilz-Sporen zu den Anilinfarben." Aerztliches Intelligenzblatt. 1884. No. 33, p. 370.

BURRILL (T. J.): "Preparing and Mounting Bacteria" Proc. Amer. Soc. Microscopists, 6th meeting, p. 79, also Micros. News, vol. IV., 1884, No. 44, p. 199.

DUFRENNE: (Improvement in photographing Bacteria). Bull Soc. Belg. Micr. X., 1884, pp. 128–32. Journ. Roy. Mic. Soc., 1884, p. 627.

EHRLICH (P.): I. "Beiträge sur Kenntnis der Anilinfärbungen." Arch. f. mikr. Anat., Bd. 13. S. 263. 1876. II. "Ueber das Methylenblau und seine klinische bakterioscopische Verwendung." Zeitschr. f. klin. Med. II. p. 70, 1881.

FLESCH: "Beleuchtungsvorrichtungen zum Mikroskopiren bei kunstlichem Lichte." Sitzber. d. phys. med. Gesellsch. zu Würzburg, 1882, p. 37.

FOL: "Lehrbuch der mikroskopischen Anatomie" Erste Lieferung. Die Mikroskopisch-anatomische Technik." p. 202–208. Leipzig (Wm. Engelmann) 1884.

FRIEDLÄNDER (C.): "Microscopische Technik, zum Gebrauch bei medicinischen und pathologisch anatomischen Untersuchungen," 2 Aufl. Berlin, 1884.

GIBBES (H.): "Double staining for Bacilli." The Lancet, 1883, p. 771.

GRAM (C.): "Ueber die isolirte Farbung der Schizomyceten in Schnitt- und Trochen-praparaten." Fortschritte der Med. Bd. 2, No. 6, p. 185, 1884.

HANSEN (E. C.): "Ueber das Zahlen mikroskopischer Gegenstande in der Botanik, " 6 woodcuts. Zeitschr. f. wiss. Mikroskopie Bd. I. H. 2, p. 191.

HAZLEWOOD (F. T.) ; (On a blue staining fluid for Bacteria). American Monthly Microscopical Journal, 1884, p. 83 ; also June, 1883.

HOFFMANN (G. v.) : " Untersuchungen über Spaltpilze im menschlichen Blute " 82 pp. 8°. mit 2 Tafln. Berlin. (Hirschwald), 1884.

HUEPPE (F.) : " Die Methoden der Bakterien-Forschung." Mit 2 Tafln in Farbendruck und 31 Holzschnitten. February, 1885. Wiesbaden, (C. W. Kreidel's Verlog).

KLEIN (E.) : " Micro-organisms and disease. An introduction into the study of specific micro-organisms." London (Macmillan), 1884, 8vo, 108 woodcuts.

KOCH (R.) : " Zur Untersuchung von pathogenen Organismen." Mittheilung a. d. kais. Gesundheitsamt, 1881. Bd. 1, p. 1. Cf. Berl. klin. Wochenschr., 1882, p. 15.

KOCH (R.) : " Verfahren zur Untersuch. z. Conserviren und Photographiren d. Bacterien." Beiträge z. Biol. Bd. II.

LEE (A. B.) : " The Microtomist's Vade Mecum. A Handbook of the Methods of Microscopic Anatomy." London (J. and A. Churchill), 1885, (Chapt. XV., " Bacteria Staining ").

MARPMANN (C.) : Die Spaltpilze. Grundzüge

der Spaltpilze oder Bactcrienkunde" (vid. parts II., III.) Halle a. S. (Waisenhauser) 1884.

MITTENZWEIG: "Ueber den Verlauf des bakteriologischen Cursus im Reichs Gesundheitsamt" Conferenz, d. med. Beamten des reg. Bez. Dusseldorf. Extr., in Deutsch. med. Wochenschr. Nov. 20, 1884, p. 769.

NOWAK (J.) : "Die Infectionskrankheiten vom aëtiologischen und hygienischen Standpunkt. Methoden der Untersuchung auf Microorganismen." Wien, 1884, pp. 142.

ORTH (J.) : "Compendium der pathologisch-anatomischen Diagnostik, nebst Anleitung zur Ausführung von obductionen von pathologisch-histologischer Untersuchungen." Dritte neu bearbeite und mit mikroskopische Technicks vermehrte Auflage. gr. 8, 1884.

ORTH : "Notizen zur Farbetechnik." Berliner klin. Wochenschr, July, 1883, p. 421.

PARIETTI (E.) : "Recerche relative alla preparazione e conservazione di Bacteri e d'Infusori" Bollett. Scientifico, vol. V., 1883, p. 95.

PERTY : "Zur Kenntniss kleinster Lebensformen," 1852, p. 13.

PEYER (A.) : "Die Microscopie am Krankenbette." 8vo. Basel, 1884, XII., 19 pp. and 79 pl.

PLAUT (H.) : "Farbungs Methoden zum Nachweis der faulniss und pathogenen Mikroorganismen," 2 Aufl. Leipzig, 1885.

RECKLINGHAUSEN : (On rendering Bacteria more visible by acids or alkalies). Verhandl. der physikal-medicin. Gesellsch in Würtzburg. N. F. II. Bd. Hft. 4, 1872. Sitzungsbericht, p. XII.

SAHLI (H.) : "Ueber die Anwendung von Boraxmethylenblau für die Untersuchung des centralen Nervensystems und für den Nachweis von Mikroorganismen, speciell zur bacteriologischen Untersuchung der nervosen Centralogane." Zeitschr. f. wiss. Mikroskopie, Bd. II. Hft. 1, p. 49.

SIEDAMGROTZKY U. HOFMEISTER : "Anleitung zur mikroskopischen und. chem. Diagnostik."

STERNBERG (G. M.) : "Photo-Micrographs, and How to Make Them." Boston (J. R. Osgood & Co.), 1883. 8vo, pp. 204.

STOWELL (C. H.) : "Bacillus Staining." The Microscope, IV., p. 79, 1884.

WALMSLEY & CO.: "Circular on Bacillus Staining." The Microscope, vol. III., p. 310 (1883); vol. IV. (1884), pp. 79, 80.

WEIGERT : "Zur Technik d. mikrosk. Bakterien Untersuch." Virchow's Archiv. Bd. 84, p. 283, 1881. Cf. Sitzung der Schlesischen Gesellschaft für vaterlandische Cultur vom 10 Dec., 1875. Berl. klin. Wochenschr, 1877. No. 18 u. 19.— Bericht über die Münchener Naturforschersversammlung, 1877, p. 283.

ZACHARIAS (O.) : " Das Mikroskopie und die

wissenschaftlichen Methoden das Untersuchungen." Leipzig, 1884.

ZOPF (W.): "Die Spaltpilze" (Abschnitt III. Methoden der Untersuchung), 2d Edit. Breslau. (Trewendt), 1884.

II. CULTURE EXPERIMENTS.

For the perfecting of this method of investigating Bacteria we are chiefly indebted to Dr. Robert Koch. In it we have a plan by which we can select from the numerous Bacteria found in a diseased body, or in a putrefying or fermenting mass, those forms which are the cause of the change. As we have already noticed, the mucous surfaces of the body are always in healthy animals the home of numerous Bacteria: in a diseased animal, how are we to find the potent factor of the morbid condition, how isolate it from the forms which are comparatively innoxious? Or, if our attention is called to some milk which has turned blue, as it sometimes does through the agency of Bacteria, how are we to select from the myriad of living forms, filaments, rods, etc., — in short, all the forms which may have entered the milk from the air of the stable or dairy, — the one which has done the mischief? That this can be done seems scarcely possible; and yet it is not so difficult, providing proper care is exercised in the processes employed.

For instance, in the case of the blue milk, we take a small quantity of the milk (what will stick to the point of a cambric needle) ; this we mix with about a teaspoonful of properly prepared gelatine, which becomes fluid at 35° C. In this melted gelatine the few microbes which we have taken up on the needle become scattered, and it is poured out upon a flat surface, properly protected from the air, and in from twenty-four to seventy-two hours we obtain from each microbe a separate pure culture,— here a mould-like patch, here a porcelain-like fleck, here a red, there a gray or blue, — all pure cultures. We may now take from each of these a particle, as we did from the milk, and make a new sowing in fresh gelatine, and by acting upon some pure, fresh milk with the various pure cultures, we find at last that those Bacteria belonging to the little bluish patch are the special agents in the case. After this plan of procedure, — but not, of course, in so simple a way, — did Koch work out the " *Comma Bacillus* " of cholera, the *Bacillus* of septicæmia in mice, and others, the specific germs of pyæmia, erysipelas, pneumonia, etc. " It is only by such *monosporus cultivations*," writes Prof. Lankester, " that we can arrive at solid conclusions in reference to the forms and activities of Bacteria, *e.g.*, as to whether one form can give. rise to progeny of another form, when its food and conditions of growth are changed, and again as to whether fermentative powers can be lost or acquired in the course of generations derived from one parent germ, but subjected to different conditions as to food, temperature, and oxygen. The method of gelatine cultivation devised by Dr. Koch places the means of following out these inquiries in the hands of every skilful microscopist."

Again, Bienstock, in connection with the statement that it is impossible to establish the identity of two micro-organisms simply from the resemblance of their morphological characters, says: "In bacterial observations, culture, and not the microscope, is the important point. The microscope is principally only an accessory checking apparatus. It gives exact information only in the study of morphology; it is an uncertain guide, and, in physiology, generally of no use at all."

Pasteur expresses himself regarding culture methods as follows: "Une méthode pour ainsi dire unique m'a servi guide dans l'étude des organismes microscopiques. Elle consiste essentiellement dans la culture de ces petits être à l'état de pureté, c'est-à-dire, dégagés de toutes les matières hétérogènes mortes ou vivantes qui les accompagnent. Par l'emploi de cette méthode, les questions les plus ardues reçoivent parfois des solutions faciles et decisives." Although many

microbes seem to flourish in solutions of inorganic salts, it is found preferable to have the medium in which the cultures are carried on rich in organic materials, containing at the same time some inorganic salts. Successful cultures depend upon the use of a completely sterilized nourishing medium and a pure sowing. These culture media may be either of a solid or liquid character, each having its own particular advantage and use. No medium can be prepared, however, which will germinate all forms of Schizomycetes indiscriminately; *e.g.*, *Micrococcus ureæ* will not germinate in broths, but in urine, while the microbes of chicken cholera will die within forty-eight hours, if placed in (l'eau de levûre) a decoction of beer yeast, which forms an excellent culture fluid for other forms, especially for filth Bacteria.

Fluid Culture Media [F. 1.13.31]. — Broths made from all kinds of flesh, from that of human beings to that of horses, have been used by various investigators; these are freed from fat, neutralized, filtered, and sterilized. Decoctions and infusions of various vegetal substances are used under special circumstances. *Sterilizing* the culture medium is accomplished ordinarily by the use of heat sufficient to kill all germs or fully developed Bacteria in the fluid. It may be accomplished in bulk, and then transferred to sterilized culture vessels, or it may be placed in these, hermetically sealed, or simply stopped with a plug of cotton, and then sterilized.

For sterilizing the culture fluid in bulk, a strong iron vessel is used, having (like a Wolff's bottle) two openings through the close-fitting top. One of these is the opening of a tube, closed at its lower end, for holding the thermometer; in the second opening a metallic tube is fitted by being tightly wrapped in cotton and pressed firmly into the opening. One end of this tube extends into the vessel, while the other is bent twice around and joined to a bit of rubber tubing, which is held closed by a wire clamp. To sterilize the culture medium, draw up the bent tube until its lower end is above the culture fluid, and then heat the whole up to 110° C. for about an hour. Now remove the clamp, and allow the steam to pass for about ten minutes. Then clamp it again, and press the bent tube into the fluid. If it is now desired to fill previously sterilized culture vessels, it may be done by relaxing the clamp, when the fluid will flow out through the tube.

Pasteur's Sterilization Method is to place the culture medium in small, pear-shaped flasks, or balloons, blown from a glass tube, which are then hermetically sealed by fusing the tube. The vessels thus prepared are sunk in a bath containing chloride of lime or nitrate of soda by means of a perforated metal plate, and are retained here for twelve hours at a temperature of 110° C. or 115° C.

Buchner's Sterilization Method requires a kettle 20 ctm. in diameter and 45 ctm. deep, which can be closed steam-tight. In the bottom of this place a layer of water 5 to 8 ctm. in depth. Then, by means of perforated diaphragms, the test tubes containing the culture fluid are arranged in two tiers, one above the other; they are closed with cotton, and each has also a bit of cloth tied over the top, and a small funnel is inverted over each to prevent any drops of water from reaching the cotton. The cover of the vessel is screwed on tight, and the whole apparatus heated for an hour and a quarter, and then kept for one hour at 110° C.

Tyndall's Method of Discontinuous Heating. Tyndall claims that it is only possible to destroy the germs, or "resting spores," of Bacteria by exposing them to alternate heat and cold, allowing in the cool intervals successive broods to develop, and killing these full-grown microbes at each heating. Koch modified this method by placing the degree to which the heating should be carried at 70° C.

Miquel's Method of Sterilizing without Heat depends upon the use of filters. For some rodlike or long, filamentous forms, paper has been found to answer; but to keep back the minute spores, cylinders of clay or plaster of Paris have been found best adapted. A flask is used, in the side of which, near the neck, a small vent-tube has been blown. The neck of the flask is also somewhat constricted at its lower end. A mixture of plaster of Paris and asbestos [For. 39] is poured into the neck of the flask, and allowed to dry slowly. The flask is then heated until the contained air has all been expelled, and any germs killed. The vent-tube is then fused together, and the culture fluid poured upon the plaster plug in the neck of the flask, through which it is gradually forced. This method is used for such culture fluids as contain albumen, which would be coagulated by the heat required to kill the germs in all the other methods except Koch's modification of Tyndall's method.

Gautier's Method of Sterilization without Heat. He employs a long-necked flask, made of "fayence" or of unglazed porcelain, and having a conically projecting bottom. Through this conical bottom the fluid is filtered from without into the flask. Fastened into the neck of the flask by means of cement (F. 44) is a glass tube bent at right angles, of which one limb reaches to the bottom of the cone, while the other outer limb tapers conically and passes into a corresponding conical expansion of a second tube. This second tube is likewise bent at right angles, and that end not united with the first tube passes to the bottom of a glass flask having a narrow neck. The side of this glass flask carries a funnel-shaped appendage. The conical openings are stopped with cotton, and the whole apparatus sterilized in separate pieces. The tube from the porcelain flask, after sterilization and the removal of the cotton, is fitted into the conical expansion of the second tube. In the funnel-shaped appendage of the glass flask is fitted a tube which is filled with sterilized asbestos. All the joints are made tight with shellac. By aspiration upon the free end of the asbestos tube, the air in the entire apparatus is rarified; and if the conical bottom of the porcelain flask is placed in a fluid, the fluid will be aspirated, free of germs, into the flask.

Koch, Gaffky, and Laeffler's Steam Sterilizing Cylinder. — This consists of a tin cylinder of about $\frac{1}{2}$ meter high and from 20 to 25 cm. in diameter, having a copper bottom, and being protected against loss of heat by an asbestos covering. At the lower third of the interior is a grate, the space beneath which is filled three-fourths full of water. This is caused to boil by from three to five gas flames beneath the cylinder. The cylinder is closed with a tin top covered with asbestos. This top does not fit hermetically, and allows of the steam escaping. Through its apex passes a thermometer. Advantage, — cheapness, and impossibility of temperature rising above 100° C.

A wire basket, fitting into the cylinder, is used for holding the tubes, etc., to be sterilized, facilitating their introduction and removal.

Solid or Gelatine Culture Media are made by adding to any of the fluid culture media sufficient (5 : 10 per cent) pure gelatine, (F. 46 e.) or, preferably, Japanese sea-weed glue, "Agar-Agar," (F. 46 d.) to give a mass which will become fluid at a low temperature, *e.g.*, 30° C.

Koch refers to solid culture media as follows (Deutsche med. Wochenschr. No. 45, 1884, p. 726): "In the ordinary method of Bacteria culture in a firm medium, the single germs must be placed as far as possible from each other, in order that they may develop entirely apart. The material containing the Bacteria is therefore placed in a gelatine nourishing medium which has been made fluid, and in this they are distributed as much as possible, and then are poured out, with the gelatine upon them, upon a glass plate; the gelatine hardening quickly, the separate Bacteria scattered through it are fixed apart from each other, and each germ can go on developing and multiplying until it becomes a mass of pure culture, visible to the naked eye, without being disturbed by or mixed with other kinds. The principle of the method is to develop entire colonies from single individuals. It is much more difficult to separate a number of different kinds of microbes on the surface of potato than by the gelatine process. In most cases the separating of pathogenic from non-pathogenic Bacteria upon potato cannot be accomplished because the Bacteria of filth grow so much more luxuriantly on potato that they soon overcome all others. Potato is therefore used for a nourishing sub-strata for pathogenic Bacteria only when these have been obtained from pure cultures and it is desired to know whether they will subsist on vegetable diet or not."

The advantage of gelatine cultures consists in the ease with which different kinds of Bacteria can be isolated and studied throughout their various phases of development. When a test-tube is used as a culture vessel for gelatine media, the mass should be allowed to cool while the tube is inclined at an angle of 45°; it will thus, when stiffened, present a much larger surface for inoculation. Thus, after the sowing instrument has been drawn over the firm surface from bottom to top, there will be a number of colonies spring up along the line of sowing, and from any one of these material may be taken for new pure cultures.

Gelatine offers a good medium for the study of the development of Bacteria, since the germs may be sown in it while fluid, and then spread in very thin layers on glass slides, which are kept under a bell-glass, and can be studied as they are with low powers, or with immersion lenses by placing a thin cover-glass over the colony to be investigated. The great objection to gelatine culture media is that they cannot be retained firm at those temperatures best adapted to the growth of Bacteria, 30°–37° C. A substitute which fulfils this requirement is *Koch's Serum Culture Medium*. (F. 46 c.) Serum of an ox or sheep is rendered as pure as possible, and placed in test-tubes closed with cotton and rubber cloth so that they are water-tight. These are for six days heated daily for one hour at 58° C., by which process the serum is in most cases completely sterilized. It is then warmed at 65° C. until it becomes stiff and firm. After this treatment the serum appears as an amber yellow, completely transparent or only weakly opalescent, firm, gelatinous mass, and after several days in the breeding oven must show no signs of developing bacteria colonies. If the heat employed in stiffening the serum is over 75° C., or lasts too long, the serum becomes opaque. To furnish a large surface for culture, the serum is allowed to stiffen while the test-tube is in an inclined position. For such cultures as are to be studied under the microscope. the serum is allowed to stiffen in a flat watch-glass or in a hollowed-out slide. Upon this stiffened blood-serum, which forms a firm nourishing medium at the temperature of the breeding oven, the material to be investigated is placed, and the whole kept at

37° or 38° C. until development takes place, which in the case of the tubercle *Bacillus* is from twelve to fourteen days. Instead of blood-serum, hydrocele fluid (F. 46 b.) may be used, being rendered solid in the same way as blood-serum; or in place of either of these, Miquel's lichen jelly (F. 46 a.), which is convenient, as it can be rendered fluid again, if necessary, which is not the case with blood-serum or hydrocele fluid.

Culture Vessels. — Ordinarily simple test-tubes closed with asbestos-wool, or with cotton are used, but if it is desirable to be especially particular, take a conical-shaped flask of good glass, with a flat bottom, also a piece of glass tubing of a diameter smaller than the neck of the flask and somewhat longer. This tube must be slightly pointed at the lower end. Pack the tube three quarters full of asbestos, and upon the top of this place a pad of cotton; wrap the tube in some cotton batting and press it firmly into the neck of the flask. This apparatus may now be heated up to 200° C. until thoroughly sterilized.

Salmon's Culture Tube. — The culture tube of Dr. D. E. Salmon consists of a test-tube-like body or reservoir, of rather heavy glass, about four or five inches in length, and three quarters of an inch internal diameter. Over the top of this reservoir a second hollow piece or cap is fitted. Its internal surface is ground to fit snugly over the ground external surface of the upper end of the reservoir, thus forming a ground joint union. This cap, about two and a half inches long, abruptly contracts near its middle into a narrow tube of about three eighths of an inch internal diameter. The third piece, or ventilating tube, is like an inverted U, one limb being about three inches long, and one and a half inches longer than the limb which fits, by means of a ground joint, over the narrow tube of the cap. The longer free limb of the ventilating tube lodges a plug of glass-wool from one and one half to two inches long. The limbs of the ventilating tube are about one inch apart. The culture fluid is introduced by removing the cap, which brings with it the ventilating tube, and it is sterilized in the tube. The liquid is inoculated by removing the ventilating tube only. To prevent the ground joints from sticking too firmly, a little sublimated vaseline is introduced between the surfaces of the joints. The pipette, used to introduce a drop of fluid containing Bacteria, consists of an ordinary glass tube about one quarter of an inch in diameter, and two or three inches long, one end of which is drawn out into a very fine, almost capillary tube, which must be long enough to easily reach the bottom of the reservoir when introduced through the narrow tube of the cap; a plug of glass-wool occupies the other end, which is closed by a rubber ball. The method of inoculating the culture liquid is briefly as follows: The pipette is first thoroughly sterilized by flaming every portion of it, from the tip of the capillary tube to near the rubber ball, until the contained air is subjected to a temperature of at least 150° C. It is usual to bring it to a dull red heat, avoiding the contingency of melting the capillary tube. It is hung with the rubber bulb up, to avoid its capillary portion coming in contact with anything while cooling. When sufficiently cool, the capillary portion is again drawn once or twice through the flame, to destroy any particles that may have become attached meanwhile. The ventilator of the culture tube containing the Bacteria to be sown is flamed and removed, and the narrow tube of the cap flamed; the rubber bulb slightly compressed and the pipette introduced, a few drops drawn up, the pipette slowly withdrawn, the cap flamed again, and the ventilator replaced. The cap of the fresh tube is now flamed before and after removing the ventilator, the pipette introduced, a drop allowed to fall into the culture fluid, the pipette removed, the narrow tube of the cap again flamed, and the ventilator replaced. When the source of the Bacteria is an exudate or the flow of the animal body, various methods are in use; the method given above may, however, be employed in most cases. The reservoir may be variously modified,—a flask-shaped body may be used for cultures that require an abundance of air,—but the test-tube form will serve nearly all purposes. It enables the nature of the opacity of the liquid to be readily determined, while the earliest traces of a membrane or deposit are more easily detected than with a broad body and a flat bottom. The culture-tube recommends itself as a simple, very neat apparatus, readily filled, sterilized and inoculated. It dispenses with the troublesome and dangerous expedient of disturbing cotton plugs, and of tying down various air-filtering materials. It is easily cleaned, and hence may be used over and over again,— the original cost of the tube being in this way reduced to a minimum in the end. It does not break readily, nor are there any sharp or jagged edges to be feared in the manipulation of dangerous cultures. It is very compact, and occupies but little space in a thermostat. Finally, the chances of contamination through the air, during the process of inoculation, are practically of no account.

Sternberg's Culture-flasks are made from glass tubing, very expeditiously, and at small expense. They are simply little bulbs blown from a glass tube, and having a long neck which tapers gradually to a capillary point. Each flask contains a sufficient amount of nutrient fluid and oxygen to ensure a vigorous and abundant development of any aerobic micro-organisms introduced as seed. The *Bacilli* readily form spores in these flasks.

When properly sterilized, the enclosed culture-medium remains closed indefinitely, and the little flasks, ready for use at a moment's notice, may be packed away in drawers or boxes for years, if desired, and may be conveniently transported from place to place. The inoculation of one flask with micro-organisms contained in another, or with a drop of blood from the veins of a living animal, etc., is effected expeditiously, and with perfect security from contamination. Small amounts of fluid may at any time be withdrawn from one of these flasks for microscopic examination, without the slightest danger of introducing foreign organisms and thus destroying the purity of the culture. Finally these little flasks take the place of a syringe, where an inoculation experiment is to be performed, the contents being forced beneath the skin of a living animal, by applying gentle heat to the ball, thus causing the enclosed air to expand, and forcing the fluid contents through the capillary neck of the flask.

To introduce the Sterilized Culture Fluid into the Sterilized Culture Vessel. For this purpose a long, sharp canula is needed, made of silver, platinum, or glass. Fasten this into the rubber tube of the vessel containing the culture fluid described above, having previously sterilized the canula by holding it in the steam escaping from the vessel, or by heating it in the alcohol flame. Remove the cotton plug from the culture vessel (in the last-described form) and push the canula through the asbestos pad into the vessel; now relax the clamp on the rubber tube, and the culture fluid will flow over into the culture vessel; after withdrawing the canula, replace the cotton.

To sow the Microbes in the Culture Medium. — Only such instruments are to be used as have been sterilized by heat. Since hot instruments may kill the Bacteria, and, if allowed to cool in the air, may be again contaminated, a plan has been recommended whereby a number of cool sterilized wires may be kept on hand (Fol.). Take an ordinary test-tube and plug it with asbestos, and pass through this plug a number of small pipettes, each of which is stopped with an asbestos plug, through which passes in each a piece of platinum wire. Now sterilize the whole tube in an oven. As required, one of these cool sterilized pipettes and wires may be removed and used. Thrust one of these wires into the tissue the microbes of which it is desired to study, draw it out, and in the meantime having thrust one of the sterilized pipettes through the asbestos pad into the culture vessel, pass the wire through it into the culture fluid, gelatine, or blood serum, as the case may be, and the sowing is completed. If the sowing is to be made from a fluid instead of a tissue, take one of the sterilized pipettes without the wire, and, placing over it a rubber cap, draw up some of the fluid and pass through the asbestos pad as before. Then replace the cotton above the asbestos.

To carry on the culture, it is best, with fluid media, to keep them at a temperature of 35° C. in a breeding oven. This is also the plan for serum cultures, but gelatine cultures must remain in the room at ordinary temperatures, at which only do they remain firm. Otherwise we should lose the advantages gained by the addition of the gelatine.

Where it is not necessary to exercise great precautions against contamination, as in growing the various pigment Bacteria, cultures may be carried on under bell-glasses by sowing the Bacteria upon slices of boiled potato, turnip, on eggs, or any similar nourishing substances, as the gelatine or Agar-Agar masses on glass slides or in watchglasses. Place a piece of moistened filter paper upon a plate, and another piece in the dome of the bell-glass; then place the vegetable, egg, or gelatine with its Bacteria on the plate, cover and keep in a moderately warm place. The colonies, especially in the case of pigment Bacteria, will soon make themselves visible to the naked eye.

Van Tieghem and Lemonnier's, and Miquel's Methods for the cultivation of Bacteria upon the slide. In the centre of an ordinary slide is fastened, by means of Canada balsam, a glass ring from 4 to 5 mm. thick, cut from a tube used for organic analysis, and the cut sides properly ground level. A thin cover-glass, round, and of a sufficient diameter to just cover the ring without lapping the edge, is fixed on the upper side by three very small drops of a greasy oil, to complete the cell. In order that the interior air may be always saturated by moisture, a few drops of water are placed on the bottom of the cell. A small drop of nutritive liquid is suspended at the centre of the under surface of the thin cover. In this drop are sown the seed microbes. Such was the moist chamber used by Van Tieghem and Lemonnier; it has, however, undergone, during the last ten years, many modifications, more or less satisfactory. Some investigators have pierced the sides of the little chamber with one or more holes for facilitating the introduction into the interior of various reagents. For the study of the atmospheric Schizophytes, Miquel uses the same cell pierced laterally by an opening which can be closed by a small glass rod stopper. The cells and cover should be attached to the slide by a cement that will not be loosened by the heat used to sterilize the chamber [F. 44]. Then, by means of a pipette with a curved capillary point, the sterilized nutritive liquid, or blood-serum, broth, urine, vegetable juices, etc., is placed upon the under surface of the thin glass cover, whilst the sowing of the organisms whose development is to be watched is accomplished by the aid of a fine platinum wire, slightly bent at the point. The small rod stopper is replaced, and the whole, with the microscope, is placed in a warm chamber at 30° C. If immersion objectives be used, a little glycerine can be added to the water, or cedar oil used on the cover. Good, dry objectives, and the light from a paraffin lamp, will generally suffice; but Miquel gives preference to the excellent No. 7 immersion objective of Nachet. It is not necessary to further describe the precautions required to prevent contamination, and the neglect of which may entirely nullify the value of the cultivation. Another form is one in which there are two opposite lateral openings into the cell; in one is fitted a tube containing a plug of cotton; through the other, air is projected upon the drop of nutritive fluid on the under side of the cover-glass, by means of a curved pipette, the tube containing the cotton serving as an aspirator. After this has been done, the second hole is stopped with a cork.

Salmonsen's Method for pure cultures. "An absolutely pure sowing was taken from putrefaction-specks in defibrinized ox-blood which had been preserved and observed in capillary tubes. In order to obtain as large a number as possible of different forms, he chose (1) those specks which presented the greatest possible differences in reference to time of incubation, rapidity of growth, and appearance; (2) specks from the blood of different individuals; and (3) he employed that blood only which contained only a small number of specks, which were therefore at a distance apart. The piece of capillary tube, the contents of which were going to be sown, was then separated under water by a strong pair of scissors, and placed in the culture-bulb with the requisite precautions — viz., using all the instruments immediately after strong heating, to destroy the dust, etc. A bulb was used for this purpose, provided with a rather short (4 cm.) and relatively wide neck, with only a small opening closed with a cap. The cap was composed of a rubber tube, which was so firmly closed for half its length by a small wad-stopper that it was slightly bulged. The tube was somewhat wider than the upper end of the neck of the bulb, that it might be placed on it without difficulty, but narrower than the lower part. After the requisite quantity of the nutrient fluid had been drawn into the bulb, and the latter closed, the definite purification and sterilization were effected by boiling." (Journal Micros. Soc., 1880, p. 649.)

LITERATURE ON CULTURE METHODS.

ALMQVIST (E.) : "Die besten Methoden Bacterien rein zu cultiviren." Botan. Centralbl. Bd. XIV. 1883, p. 286.

BIEDERT: "Platten Culturen," Deutsche med. Zeitung, 1884.

BREFELD: I. "Kulturmethoden zur Untersuchung der Pilze," Botanische Untersuchungen über Schimmelpilze. Hft. IV. (1881). II. "Die kunstlicher Kultur parasitischer Pilze." Botan. Unters. über Hefenpilze. Bd. V. 1883.

ERMENGEN (VAN): " Sur les méthodes de culture des micro-organismes pathogènes." Bull. Soc. Belg. de Microsc. t. IX., 1883, No. 8, p. 105.

FELEISEN: "Ueber neue Methoden d. Untersuchung und Cultur pathogener Bacterien." Phys. med. Gesellsch. zu. Würzburg, 1882, p. 113–121.

HEYDENREICH (L.): " Sur la sterilization des liquids au moyen de la marmite de Papin." Compt. Rend. T. 98, p. 998 (1884).

JOHNE: "Ueber die Koch'schen Reinculturen." 1885.

KLEBS (E.): " Ueber fractionirte Cultur " in Archiv. f. experim. Pathol. Bd. I, p. 46, also Sitzber. Phys. med. Gesellsch. zu Würzburg, 1873.

KOCH (R.): "Platten kulturen." Vertrag auf dem XI. deutschen Aerztetage, 1883, zu Berlin.

PASTEUR (L.): (Methods of culture in the study of alcoholic fermentation). Annales de Chemie et de Physique. vol. 58, p. 323.

SALMON: (On a New Form of Culture Tube). American Monthly Microscopical Journal, V. (1884). — Journal Roy. Mic. Soc., 1885, p. 145.

SALMONSEN: "Zur Isolation differenter Bakterien formen." Bot. Zeitung, 1876, No. 39. "Studier over Blodets Forraadnelse," 1877. "Eine einfache Methode zur Reinkulturen verschiedener Faulnissbakterien." Botan. Zeitung, 1880. No. 28.

SMITH (T.): "Remarks on fluid and gelatinous media for cultivating micro-organisms, with description of Salmon's new culture tube, and demonstration of the process of using it." Amer. Monthly Mic. Journ. V. (1884), p. 185.

STERNBERG (G. M.): "Methods of cultivating micro-organisms." Report Amer. Assoc. Adv. Sci., 1881. American Monthly Microsc. Journ. V. (1884), pp. 183–185.

III. VACCINATION OR INOCULATION EXPERIMENTS.

In inoculating animals with the pure cultures of pathogenic Bacteria, it is found that only particular animals are susceptible to the influence of certain microbes; thus birds, especially birds of prey, are not affected by the *Bacilli* of anthrax; the same little rods which produce a fatal disease in the domestic mouse are harmless to birds, rabbits, etc. Even animals of the same family are exempt; thus, in the latter case, the field mouse is unharmed by the microbes which prove fatal to the domestic mouse. The mucous surfaces of mammals, and serous surfaces as well of lower vertebrates and invertebrates, are not suitable places for the introduction of special germs whose action on the system it is desired to investigate, inasmuch as these surfaces are invariably inhabited by a great variety of microbes, as we have already noticed. The serous surfaces of birds and mammals, however, are not in normal conditions, so far as is known, inhabited by any adult microbes.

We have, therefore, two plans by which the vaccination or inoculation from pure cultures may be successfully made. After having removed all hair or feathers from the selected spot, and having thoroughly washed the skin, and bathed it with an antiseptic solution (*e. g.*, two per cent corrosive sublimate solution), dry it with absorbent cotton, and make a cut with a previously sterilized knife; into this introduce the microbes from the pure culture by means of the sterilized platinum wire. If the spot is now covered with some surgeon's gauze or cotton, all inoculation from the air may be avoided. The pure culture, in case of fluids, may, by means of a sharp, sterilized canula, be injected at once into the mesodermal tissues; an ordinary hypodermic syringe, when properly sterilized, answers the purpose. Any special modifications of the general plan will be noticed in the appropriate place under the treatment of pathogenic Bacteria.

Inoculation methods must occupy a subordinate position until the subject of attenuation is better understood. A number of years since, Virchow (Archiv für patholog. Anat. Phys. Bd. 79, p. 213) made the following remark: " Truly, we know nothing regarding it (the question of immunity), and the knowledge of Bacteria has also not shed the least light upon the question." That we now are in scarcely better position is indicated by the following, taken from a recent article by so distinguished a physiologist as Lauder-Brunton ("Practitioner" for September, 1884). "To sum up the results of these laborious investigations, we must agree with Koch and Klein that immunity can be conferred by inoculation, but this immunity can only take rank as an interesting theoretical fact, not yet brought within the region of practical science. We must deny Pasteur's claim that his method of animal inoculation gives absolute immunity, and that it is harmless, the number of fatal results being nil or very small. There is no doubt that when Pasteur performs his inoculations without any deaths, he is working with cultures too weak to give any immunity, and also that in many cases the 'virulent' anthrax *Bacilli*, sent out by him as a test for the protection already conferred by vaccination, are much too weak. The 'vaccins,' as supplied by him, have been found very variable; thus, sometimes his '*vaccin premier*' has killed a flock of sheep, whilst '*vaccin deuxième*' has been inert. The objection to the present method is that immunity can only be conferred by a percentage of losses from death greater than would result if the flock were turned upon a notoriously infected pasture, and then, further, this inoculation favors the spread of the disease by the formation of spores, when any of the *Bacilli* fall on the wool of the animal; also, the immunity thus given lasts, at the most favorable computation, no longer than a season. That it may be possible in the future to discover a method of attaining immunity without too great a loss during the process must be allowed; but to consider it proved that we at present possess such a means, in the method of inoculation described by M. Pasteur, can only lead to disappointment, and to a monetary loss on the part of those adopting it."

LITERATURE ON INOCULATION METHODS.

GRAWITZ (P.): " Die Theorie der Schutzimpfung;" Virchow's Archiv., Bd. 84, 1883, p. 87.

HAY (M.): " Die Technik der Vaccination mit animaler Lymphe," Wien: 1881.

KLEIN (E.): " Micro-organisms and Disease." "Practitioner," London, Oct., 1884, p. 248.

SEMMER (E.) UND RAUPACH (C.): " Beiträge zur Lehre von der Immunität und Mitigation." Dtsch. Zeitschr. f. Thiermedicin, etc., VII., p. 347–363.

STERNBERG (G. M.): " Vital resistance theory of disease." Amer. Journal of the Med. Sci., Apr. 1,

1881. II. "What is the explanation of acquired immunity from infectious disease." (Abstr. of a paper read before the University Scientific Association, Dec. 3, 1884.) Johns Hopkins University Circular for January, 1885, vol. IV. No. 36, p. 31.

IV. BIOLOGICAL ANALYSIS.

By this method, answers are given to the numerous questions which present themselves regarding the various phenomena in the life history of Bacteria. The results of investigation so far tend to contradict the generally accepted idea that Bacteria are the simplest of all living organisms. Engelmann says in this connection: "Any one who has closely studied the form and mode of living of even the most common forms of Bacteria cannot doubt, as I do not doubt, that these organisms are relatively highly organized, and can only be considered as the simplest of organisms because they are the smallest which we can know; but small and simple are two different things. . . . Morphologically considered, they can no longer be said to belong to the simplest organisms, every Moner, every Plasmodium, stands lower. . . . I have often watched the separating of the active bacterial forms from the stiff *Cladothrix* tree; it is a wonderful sight, and I do not know with what to compare it: this plant at the moment of its change into an animal. Physiologically considered, one must grant that they possess, like true animals, certain powers and senses, e. g., the power of distinguishing between oxygen and carbonic acid, in short, a capability of breathing; and further, when one sees how they swarm about, and consume a mass of nutriment in the drop, one must grant to them, in addition, another sense, i. e., the need of nourishment. There are also Bacteria which possess a specific, and, indeed, a highly organized sense for light and color. Of course, by these phenomena, they are elevated far above the lowest organisms, above all typical plants; they are, physiologically considered, as far as they show the above and similar conditions, undoubtedly animals, and it is only so much the more powerful proof of the unity of nature that they, through their morphological development, extend into the kingdom of plants."

Zopf, Koch, and others indicate, among the questions to be answered in a thorough study of any peculiar Bacteria, the following : —

1. Shape, size, color, and details of structure, e. g., flagella, peculiar envelope, etc. ?

Character and speed of movements?

2. Character of natural habitat?

Artificial media best adapted to growth and reproduction?

Stages of development passed through?

Formation of zoogloea, spores, filaments, rods, cocci, "swarms?"

Conditions under which such formation occurs?

Character of colonies formed in firm culture media?

3. Capability of producing fermentation? putrefaction? Character of decomposition products, volatile and other, formed in various nourishing media?

4. Behavior towards oxygen at normal and altered pressures?

Behavior towards other gases?

5. Effects of various temperatures on movement, germination, etc.?

6. Behavior in relation to light (phototonic properties)?

Behavior towards electricity?

7. Behavior towards antiseptics and poisons?

8. Are the forms under investigation found in a diseased organ or tissue? What is the effect of inoculating animals of different orders and species with pure cultures?

If virulent, can the virulence be attenuated by exposure to air, to antiseptics, to heat, or by repeated "fractional" cultures? Under what conditions?

Does the inoculation of attenuated germs have a cumulative effect if repeated at short intervals?

Does one inoculation give immunity towards a second made with virulent microbes?

Bacteria may be counted and measured by means of Professor Thoma's apparatus for counting blood corpuscles, as made by Zeiss, or with an ordinary eye-piece micrometer. Szpilmann's method for studying the effect of gases was by the use of a Recklinghausen's moist chamber, with which he connected two tubes, one admitting the gas, the other allowing of its escape. The temperature was maintained at the desired point by means of a Valentine's warm table. In place of the above, a Ranvier's gas-slide may be used.

Engelmann's Method for studying the effect of light on certain Bacteria. In studying the effect of light, Engelmann used a Sugg's lamp, which he placed in a dark box [1] having but one round opening; by removing all diaphragms he could throw an image of the lighted opening at any point in the field, and in the case of *Bacterium photometricum* found that within a few minutes all the microbes would assemble in a dense mass at the lighted point. By means of the thick layer obtained in this way, he was able to subject them to spectroscopic examination, to determine their color, and to show that the coloring material had none of the peculiar characters of chromophyll. He found that without light there was no movement, and that with the return

---

[1] For a detailed description of this "Mikroskopirkasten," vid. Pflüger's Archiv. f. Physiol. Bd. XXIII., p. 577 (1880).

of light movement began anew, a condition to be compared to the phototonus of higher plants. Certain other Bacteria, e. g. *Spirillum* and *Bacterium termo*, have such an affinity for oxygen that they may be used as agents for detecting traces of free oxygen at any point of a microscopic preparation, and by means of this Engelmann was able to observe, through the working of different parts of the spectrum, the decomposition of carbonic acid and the liberation of oxygen by the organisms (in the case of *Bacterium chlorinum*, the coloring matter of which, unlike that of *Bacterium photometricum*, contains chlorophyll).

Engelmann's Method for the determination of a chromophyll assimilating power in any given Bacteria by means of other Bacteria used as reagents for the detection of free oxygen. To a drop containing as large a number as possible of the Bacteria to be examined, add a very small quantity of a fluid which contains numerous specimens of good, lively *Bacterium termo* or *Spirillum* (the latter are more susceptible reagents for oxygen than the ordinary filth Bacteria, *B. termo*), used as reagents for the oxygen. Seal up the preparation with some vaseline. After ten or fifteen minutes all or most of the individuals of *B. termo* or *Spirillum* will come to a rest from a lack of oxygen, if none is given out by the Bacteria under investigation.

Examine points where the latter have assembled in large masses, and see if the reagent Bacteria have gathered about these masses and there retain their movements, as they do around an air bubble or a green cell.

The effects of antiseptics and poisons upon Bacteria are to be studied by adding to given amounts of culture fluids containing the living microbes, certain amounts of antiseptic solutions of a known strength, varying the latter in a regular ratio through a series of experiments. Electricity and heat may be applied by means of the numerous simple contrivances described in every manual of histology and optician's catalogue.

Duclaux's Method for studying the effect of sunlight upon the germs of Bacteria. The experiments were made with *Tyrothrix scaber*, a form which flourishes well in cultures of milk and Liebig's broth. A small drop of a pure culture was taken at the moment of the formation of spores, and placed in the bottom of a small culture flask or balloon, closed with sterilized cotton, which allows free access of air, but not of new germs. The drop evaporates, the flask being exposed to the summer sun upon a wall for fifteen days, one month, or two months. Other flasks prepared in the same way were kept in an oven, where they had diffused light and a temperature about that of the maximum obtained from the sun. At the end of the periods named, culture fluids were added to the dried germs, and it was found that the germs which had been exposed to the sunlight had lost their vitality, while those exposed to a like temperature, but deprived of sunlight, reproduced and multiplied very rapidly.

Pictet and Yung's Method for ascertaining the action of cold upon microbes. The various cultures containing numerous lively microbes were hermetically sealed up in tubes, which were placed in a wooden box enveloped by such substances as are poor conductors of heat, and there kept at $-70°$ C. by the evaporation of liquid sulphurous acid, or by means of solid carbonic acid, the latter being constantly renewed by means of tubes without altering the pressure. The microbes were kept at $-70°$ to $-76°$ C. for forty-eight hours, allowed to stand for six hours at ordinary temperatures, and then again frozen. The degree of temperature was determined by means of the formula given by MM. Raoul Pictet, and Cellerier in their Memoire upon the maxima tensions of saturated vapors.

LITERATURE OF THE BIOLOGY OF BACTERIA.

BÉCHAMP (A.) : " Les Microzymas gastriques et la pepsine." Compt. Rend. T. XCIV., p. 970.

BLACK (G. V.) : " The formation of poisons by micro-organisms, a biological study of the germ theory of disease." Philadelphia Blakiston), 1884, 12mo.

BOEHLENDORFF (H. v.) : "Ein Beitrag zur Biologie einiger Schizomyceten." Inaug. Dissert. Dorpat, 1880.

BRIEGER (L.) : "Ueber Spaltungsproducte der Bakterien.' Zeitschr. f. phys. Chemie, VIII., p. 306.

BRISSON (T.) : (Fatty bodies as generators of Bacteria). Untersuch. über niedere Pilze aus dem pflanzen physiol. Institut. München, I. (1882), p. 129–159.—Journ. Roy. Mic. Soc., ser. II., vol. III., p. 106.

BUCHOLTZ (L.) : I. "Antiseptica und Bacterien." Archiv. f. experim. Pathol. Bd. IV. 1875. II. "Ueber das Verhalten von Bacterien zu einige Antisepticis. Inaug. Dissert." Dorpat, 1876.

CERTES (A.) : "De l'action des hautes pressions sur le phénomènes de la putrefaction et sur la vitalité des micro-organismes d'eau douce et d'eau de mer." Compt. Rend. T. 99. 25 Aug. 1884. p. 385.

CHAIRY : "Action des agents chimique puissantes sur les bacteries du genre Tyrothrix et leur spores." Ibid. T. 99. Dec. 1, 1884. p. 980.

CHAPPUIS (E.) : (Action of ozone on germs contained in the air). Bull. Soc. Chim. XXXV., p. 390.—Journ. Chem. Soc. (abstr.), XI. (1881),

p. 632. Journ. Roy. Mic. Soc. (ii), vol. I., p. 781.

DE LA CROIX (N. J.) : "Das Verhalten der Bakterien des Fleischwassers gegen einige Antiseptica." Archiv. f. experim. Pathol. Bd. XIII. H. 3. p. 175.

DUCLAUX (E.) : "Influence de la lumière du soleil sur la vitalité des germes de microbes." Compt. Rend. T. 100. No. 2. Jan. 12, 1885. p. 119.

ENGELMANN (TH. W.) : I. "Untersuchungen über d. quantität Bez. zw. Absorption des Lichtes und Assimilation (microspektral Photometer)." Botan. Zeitung, No. 6, 7. 1884. II. "Zur Biologie der Schizomyceten." Pflüger's Archiv. f. Physiol. Bd. 26. p. 537. III. "*Bacterium photometricum.*" Ibid. Bd. 30. p. 95. 1882. Taf. 1. IV. (Describes his dark 'Mikroskopirkasten') ibid. (1880), p. 577. V. (Bacterial investigation of sunlight, gaslight, and the light of Edison's lamp.) K. Accad. Wetensch. Amsterdam, Nov. 25, 1882.—Botan. Centralbl. XIII. (1883), p. 214. Journ. Roy. Mic. Soc., Ser. ii., vol. III., p. 401. VI. "Neue Methode zur Untersuchung der Sauerstoffansscheidung pflanzliche und thierischer Organismen." Bot. Zeitung, 1881, No. 28. VII. "Farbe und Assimilation." Ibid. 1883. No. 1. p. 2.

FORBES (S. A.) : "The use of contagious

germes as insecticides." Amer. Naturalist, 1883, p. 1169.

GAUTIER: "Sterilization à froid des liquids fermentiscibles." Bulletin de la Société Chemique, 1884, vol. 42, p. 146.

GAYON (U.): (Crystallizable substances produced by a bacterium). Bull. Soc. Bot. France, XXVIII., p. 321.

GROHMANN (W.): "Ueber die Einwirkung des zellenfreien Blutplasma auf einige pflanzliche Mikroorganismen" (Schimmel — spross — pathogene u. nicht pathogene Spaltpilze), Dorpat. 1885 (C. Kruger), gr. 8.

GROSSMANN U. MAYERHAUSEN: (Effects of Oxygen and Hydrogen on Bacteria). Pflüger's Archiv. Bd. XV., 1877. p. 245.

HABERKORN (TH.): "Das Verhalten von Harnbakterien gegen einige antiseptica." Inaug. Dissert. Dorpat, 1879.

HAMLET (WM.): "Action of Compounds Inimical to Bacterial Life." Journ. Chem. Soc., XXXIX. (1881), p. 326–331.

HOPPE-SEYLER (F.): (Action of Oxygen on Low Organisms). Zeitschr. f. physiol. Chem. VIII. (1884), p. 214. — Naturforscher, XVII. (1884), p. 116.

HUEPPE: "Ueber einige Vorfragen zur Desinfectionslehre und über die Hitze als Desinfectionsmittle." Deutsche-militärärztl. Zeitschr. 1882. No. 3.

JAMIESON (J.) : "The influence of light on the development of Bacteria." Nature, 1881. vol. XXV.

KOCH, GAFFKY, AND LŒFFLER : "Versuche über die Verwerthbarkeit heisser Wasserdämpfe zu Desinfectionszwecken." Mitt. a. d. kais. Gesundheitsamt. I., 1881, p. 322.

KUHN (P.) : "Ein Beitrag zur Biologie der Bakterien." Inaug. Dissert. Dorpat. 1879.

LUDWIG (F.) : "Ueber die spectroscopische Untersuchung photogener Pilze (*Micrococcus pflügeri*)." Zeitschr. f. wiss. Microsc. Bd. I. H. 2. p. 181.

MELSENS ; "Reclamation de priorité, à propos de communications recents sur la vitalité des virus et de la levure de bière." Compt. Rend. T. 98, p. 923.

MEYER (VAN OVERBECK DE —) : "Ueber den lähmenden Einfluss von Sauerstoff sehr hoher Spannung auf Bacterien" in das Onderzoekingen ged. in het. physiol. labor. Utrecht. Derde. R. VI., 1881, p. 151–196.

MIQUEL ET BENOIST : "Sterilization à froid." Bull. de la Soc. Chem. de Paris (1881), vol. 35, p. 552.

MIQUEL (P.) : "Les organismes vivant de l'atmosphère." Paris, 1883. Chapt. IX. "Des substances Antiseptiques."

NAEGELI (C. v.) : I. "Untersuchungen über nieder Pilze," aus dem pflanzen-physiolog. Inst. in München. (R. Oldenberg.) 1882. II. "Ueber die Bewegungen kleinster Körperchen." München Acad. Sitzungsber, math-phys. Kl. 1879, p. 389. III. "Ernährung der niederen Pilze durch Kohlenstoff und Stickstoffverbindungen." Untersuchungen über niedere Pilze, 1882.

NENCKI UND LACHOWICZ: "Die Anærobiosefrage." Archiv f. d. ges. Physiologie. Bd. XXXIII. (1883).

NENCKI U. SCHAFFER: "Ueber die chemische Zusammensetzung der Faulnissbacterien." Journ. f. pract. Chem. N. F. XX., p. 443.

PATOUILLARD (N.): (Phosphorescence caused by Bacteria). Revue Mycol. IV. (1882), p. 208–209, 1 pl.—Journ. Roy. Mic. Soc., Ser. II., vol. iii., p. 106.

NENCKI (M.): "Beitr ge zur Biologie der Spaltpilze," mit 2, lith. Tafl. Leipzig. 1880.

PICTET ET YUNG; "De l'action du froid sur les microbes." Compt. Rend. T. 98, p. 467. 1884.

REINKE (J.): (Influence of Concussion on the growth of Bacteria). Pflüger's Archiv. f. Physiol. XXIII., p. 434. Naturforscher, XIV., 1881, p. 56.

RICHET (C.): "De l'action toxique comparée des métaux sur les microbes." Compt. Rend. T. XCVII. (1883), p. 1004–1006. cf. Journ. Roy. Mic. Soc., 1884, p. 427.

SCHNETZLER (J. B.) : "De l'action du Curare sur les fibres musculaires, les cils vibratils et les Bacteries, contribution à l'étude des Bacteries." Bull. Soc. Vaudois Sci. Nat. XVII. (1881), p. 625–632.

SCHWARTZ (N.) : " Ueber das Verhalten einiger Antiseptica zu Tabacksinfusbacterien." Pharmaceutische Zeitschr. f. Russland, 1880.

THIN (G.) : "On the absorption of pigment by Bacteria." Proc. Roy. Soc., London, vol. 31 (1881), p. 503–504.

TIEGHEM (P. VAN): (Bacteria living at high temperatures). Bull. Soc. Bot. de France, XXVIII. (1881), p. 35, 36.

VANDEVELDE (G.) : "Studien zur Chemie des *Bacillus subtilis*." Zeitschr. f. phys. Chemie. VIII., p. 367.

WOSNESSENSKI (J.) : "Influence de l'oxygène sous pression augmentée sur la culture du Bacillus Anthracis." Compt. Rend. T. 98, p. 314. (1884.)

WERNICH (A.) : " Die aromatischen Faulnissproducte in ihre Einwirkung auf Spalt- und Sprosspilze." Virchow's Archiv. f. path. Anat., etc. Bd. 78. p. 54.

WORTMANN (J.) : (Diastatic Ferment of Bacteria). Zeitschr. f. physiol. Chemie. VI., p. 287. Naturforscher, XV. (1882), p. 321.

BACTERIA INVESTIGATION.

PART II.

SPECIAL METHODS FOR INVESTIGATING PARTHOGENIC BACTERIA.

ANY discussion as to the rôle played by Bacteria in disease is here out of place : whether they are simple concomitants, the Bacteria of health taking advantage of a lowered vitality to develop with unwonted vigor and rapidity, thus abetting the disease by depriving the tissues of the materials required for their support (*e.g.*, oxygen, pabulum), as indicated by Pasteur for the microbe of chicken cholera : "On peut aisèment le comprendre. Le microbe, par example, est aérobie ; il absorbe pendant sa vie de grandes quantities d'oxygéne, et il brûle boucoup des princeps de son milieu de culture, ce dont il est facile de s'assurer en comparent les extraits du bouillon de poule avant et apres la culture." Whether they bring about death or local disturbance by interfering mechanically with the vital functions as emboli, or whether they are simple carriers of the infectious material, alcaloid, zymase or ptomaine, or produce them by their activity or by chemical changes which occur after they have finished their life history, as Dr. Freire thinks is the case in yellow fever, whether they are in themselves the specific *materies morbi* or not, are questions to which we can only refer in passing to the real intent of this little book, which is to place before investigators the means of answering these questions for themselves. That Bacteria are present in disease, as well as in health, cannot be disputed; that they often seem to be the cause of death, if not the exciting cause of the disease, is frequently remarked. By the following pages it will be seen that many specific forms have been described. The reasons for doing so are in some cases apparently well founded, while others need to be further investigated, and more testimony brought to bear. In this connection Dr. Koch briefly characterizes his position as follows: —

"It is not yet proven that all infectious diseases arise through parasitic micro-organisms, and on this account the parasitic character of the disease must in every case be proven. The first step to this knowledge rests upon the careful investigation of all portions of the body altered by the disease, to render certain the presence of parasites, their distribution in the diseased organs, and their relations to the tissues of the body. It is apparent that all helps which are offered by modern microscopical technology should be brought into use. The tissues and the tissue juices, the blood, lymph, etc., fresh, with and without reagents, are to be examined microscopically. They are dried upon the cover-glass, and handled according to the different processes of staining. The hardened objects can be cut into sections by means of the microtome, then stained, and the microscopic preparations thus obtained subjected to a penetrating examination by means of suitable illumination and with the best objectives. Only after a thorough investigation has been made in this way, to ascertain whether micro-organisms are to be found in diseased parts, in what organ they are to be found in the greatest purity, whether, for example, in the lung, spleen, blood, etc., can an effort be made to arrive at a conclusion as to whether they are of a pathogenic nature and the special cause of the existing disease. For this purpose they are then raised in pure cultures, and when they are by this means freed from all the originally adhering elements of the diseased body, they are inoculated back on the same species of animal if possible, or on some animal which will show unmistakably the same symptoms. In order to illustrate this, I recall tuberculosis. First, it is established, by microscopical investigation with the aid of staining reagents, that there are strongly characteristic *Bacilli* present in the diseased organs; these *Bacilli* are then isolated in pure cultures, in such a manner that they are not mixed with other *Bacilli* and thereby made impure, and are next inoculated back on as many animals as possible of such different kinds as are known to be susceptible to this disease. The disease is engendered anew. Another very good illustration is erysipelas of man. It has been long known that *Micrococci* are constantly to be found in the lymph vessels of the skin in this disease; this, however, did not prove by far that they were the cause of the disease. But since Fehleisen has succeeded in making cultures from pieces of skin cut from a person suffering from erysipelas, purifying from the possible presence of other germs on the surface by cauterization, and after breeding the *Micrococci* in pure cultures, producing a typical erysipelas again upon other men through inoculation, there can be no further doubt that *Micrococci* are the cause of erysipelas, and it is to be regarded as a parasitic disease."

The following remarks of M. Pasteur may not be out of place just here, suggesting as they do the propriety of carefully avoiding any unnecessary additions to the specific nomenclature of Bacteria. " Qu'il me soit permis d'ajouter que, dans mes Communications concernant les organismes microscopiques, je me suis abstenu généralement de donner des noms specifiques à ceúx de ces organismes que je pouvais croire nouveaux. Si cela etait nécessaire, il est toujours préférable de caractériser ces petits êtres par une ou plusieurs de leurs fonctions. Autant les dénominations speciales sont utile et commodes quand on les appliques à des êtres bien connus, autant elles peuvent créer d'embarras et de confusion lorsqu'il s'agit d'organismes tres-voisins par leurs formes et qui peuvent être tres-dissemblables par leurs propriétés physiologiques."[1]

ANTHRAX.
(Malignant pustule, Splenic fever.) *Bacillus anthracis*
(Cohn).

Toussaint's Method. — In 1880, M. Toussaint, Professor in the Veterinary School in Toulouse, announced to the Academy of Sciences of Paris that he had discovered a process by which he could, by inoculating animals with an artificially weakened anthrax poison, give them an entire immunity towards further effects of this poison. His method of reducing the virulence of the poison was — (1) to place some of the defibrinated blood of an animal suffering from anthrax, for about ten minutes, under the influence of a temperature of 55° C., using it, after cooling, for inoculation ; or (2) to treat anthrax blood with about one quarter per cent of carbolic acid, and use this mixture for inoculation. He believed that by these methods the *Bacilli* were killed, without at the same time destroying the infectiousness of the material con-

---

[1] Compt. Rend. T. 88 (1879), p. 1217.

cerned, but only rendering it milder. Loeffler, however, proved this idea to be fallacious, and experiments were made by Bouley, Chaveau, and others, but it was not until Feb. 28, 1881, that the proper method of treating the *Bacilli* was made public.

At this time the Academy of Paris received the announcement of —

Pasteur's Method, by which he was able, through artificial culture, to produce a lymph which could be used without danger as a protective vaccination against anthrax. This weakening of the virulence of the microbes was brought about by exposing them to the influence of atmospheric air under certain conditions, and he had successfully vaccinated sheep, rabbits, and guinea-pigs. He found by experiment that anthrax *Bacilli* develop best at a temperature of from 25° to 40° C.; at higher or lower temperatures the increase is lessened, stopping entirely at 15° C. as well as at 45° C.

It may here be briefly explained that the anthrax *Bacilli* possess three developmental forms —(1) little rods (*Bacteridia*); (2), spherical, spore-like bodies, called "resting spores;" and (3), minute glistening granules (Virchow and Ruloff). In the "resting" spore stage they exhibit a great power of resistance towards external influences. Pasteur investigated anthrax *Bacilli* contained in blood, at a temperature of 42°–43° C., in a culture fluid of beef broth. He studied the effect of exposing them to the influence of filtered atmospheric air; this he accomplished simply by stuffing both openings to the culture apparatus with cotton, allowing the gases of the atmosphere to pass, but excluding any contained organic germs. He found that under these conditions the *Bacilli* lost their virulence to such an extent that he could in the course of fourteen days vaccinate sheep without danger. In cases where the culture had only remained for twelve days, one half of the animals vaccinated died. In the cultures held at 42° : 43° C. the *Bacilli* retained their vitality for from four to six weeks, and were useful for vaccination during that time; after that, they died. He found that if sheep which had been vaccinated with a culture fluid, twenty-four days old, containing living anthrax *Bacilli*, were again vaccinated within twelve days with *Bacilli* taken direct, and in their full virulence, from a sick animal, they died; but if this second vaccination was made with a twelve days' old culture, the sickness following was slight, and the sheep were afterwards found to be fully protected against vaccination with the most virulent germs. The second vaccination should not, in any case, be made before twelve days have elapsed, as the effect seems to be cumulative, and death will follow in many cases. Time enough must be allowed for the first sickness to pass off. Pasteur next proceeded as follows to make a culture which should be good for more than six weeks. Take from the culture, during the first six weeks, a drop, and convey it to sterilized meat broth which is maintained at a constant temperature of 35° C. The *Bacilli* will increase rapidly here, and the newly-resulting generation will possess exactly the same grade of virulence which was possessed by the organisms of the first culture at the moment the drop was removed. After forty-eight hours the second culture begins to form "resting spores," and becomes entirely in this condition in the course of a few days. These anthrax spores now possess the same degree of virulence as the rods at the time of removing the drop. In this manner anthrax cultures may be made of any chosen virulence, which, in hermetically sealed glass tubes, will retain their efficiency for a year or more. A culture fluid of this kind (anthrax lymph) can be used as a vaccinating material, or for making new cultures in fresh sterilized broth. Pasteur named the twenty-four day old culture, as well as the secondary brood made from it, the *first vaccine* ("premier vaccin," ersten Impstoff), because he used it for his first protective vaccination. It is comparatively harmless in large amounts for rabbits, guinea-pigs, sheep, and other larger domestic animals, while it will still kill mice.

The twelve-day old culture, and that of its secondary brood, Pasteur called the *second vaccine* ("deuxième vaccin," zweiten Impstoff), because it served him for a second vaccination twelve days after the first. Twelve days after this last, there should be, as a rule, absolute immunity towards anthrax. Pasteur and his assistants, Chamberland and Roux, usually vaccinate animals upon the ear or on the thinly-haired inner surface of the thigh, or inject the culture under the skin with a hypodermic syringe. Cattle are found to be less easily protected by vaccination than sheep. In eighty per cent of the latter, immunity is known to extend over a year, and perhaps for a lifetime.

Chaveau's Method. Toussaint's method was to attenuate the virulence of the anthrax microbes by heat applied to the germs while in the fluids of the diseased animal. Chaveau made an improvement upon this, and also demonstrated that the presence of oxygen, as supposed by Pasteur, has nothing to do with the attenuation. He placed the *Bacilli* under conditions where they were more susceptible to heat. He inoculated sterilized broth with fresh anthrax blood, placed the flask in an oven, and maintained the temperature at $42°-43°$ C., but instead of keeping the culture here, as in Pasteur's method of attenuation, for twelve or thirteen days, he removed the flask after about twenty hours, and placed it in another oven, at $47°$ C.; for one, two, three, or four hours. This terminated the process; it did not destroy the virulent agents, but they were found to have lost more or less of their noxious property. He treated cultures which had had all air removed by means of a force-pump, with equally favorable results.

Chaveau's Method for the preparation and attenuation of large cultures. Two periods are required: first, for the preparation and development of the attenuated germs; second, for the complementary attenuation of the spores which result. (1.) A drop of fresh infectious blood is taken from a diseased animal and placed in a glass culture-flask, containing twenty grammes of sterilized broth, and retained for two hours at 43° C., and then heated for three hours at 47°–49° C. (2.) Glass chemical flasks, having three openings and a capacity of one or two litres, will furnish sufficient culture virus for the inoculation of from four to eight thousand sheep. Fill the flasks five sixths full of sterilized broth. The middle opening is furnished with a long tube, descending to the bottom of the flask. It is by means of this tube, whose exterior extremity is plugged with cotton, that air is introduced in fine bubbles. Of the two lateral openings, one gives rise to an adductor tube; the other is drawn out into a slender tubule for emptying the small flasks. It is by means of this last slender tube, that, aspirating on the second tube, we introduce the prepared microbes, in proportion of one drop to ten grammes of culture fluid, or eight grammes of seed fluid to a culture of sixteen hundred grammes, which may be doubled or tripled if the seed is poor. After the introduction of the seed fluid, the slender tube is closed by fusion in the lamp flame. The large culture thus prepared is placed in a thermostat at 35°–37° C. The development is incomplete if it is allowed to remain at rest; but when air is caused to pass through it, by means of the aspirator, it proliferates abundantly. In one week the evolution is usually terminated, and a rich formation of spores is found, attenuated by the heat. The culture fluid best adapted, in Chaveau's opinion, is chicken broth (one part meat, four to five parts water). The current of air should be very regular, and traverse the culture fluid in quantities of about one or one and a half litres per hour. Shake the flasks carefully, night and morning. The nearer the temperature is kept to 40° C. the better the culture. (3.) From a large flask fill a dozen of the little tubes used by Pasteur for the distribution of vaccine, and place some of these in a water-bath, others in an air-bath, and heat carefully up to 89°–90° C. Watch carefully, and note the degree of temperature at which all power of proliferation is destroyed. Now for the vaccine to be used first ("premier vaccin") heat the cultures to the nearest possible point to that which deprives them of proliferating power; for the second vaccine ("deuxième vaccin") heat to a point two degrees less than the first. Each time a large culture is made, the degree of this final heating must be ascertained; usually 84° C. answers for the first, and 82° C. for the second. Sometimes 80° C. has given a good première vaccin, and 78° C. a second. Cultures placed under the same conditions will not always give the same degree of attenuation in their virulence. Use a large amount of water in the water-bath, and have the regulator so arranged that the heat does not vary from the required temperature.

Chamberland and Roux's Method. To overcome the objection to Chaveau's method, that the cultures made with his attenuated microbes did not retain their attenuation, Chamberland and Roux proceeded to modify Toussaint's second method. They introduced into beef broth cultures of *Bacillus anthracis* which had been neutralized with potash, variable quantities of antiseptics, and placed the cultures in an oven at 35° C. The flocculent growth of the Bacteria was greatest where the least antiseptic had been used, gradually diminishing as the percentage of antiseptic increased up to a certain proportion, at which life was not manifest. Cultures made from their attenuated microbes retained their attenuation. According to these investigators, the essential condition for attenuating the virulence of *Bacillus anthracis*, whether by the methods of cultures at 42°–43° C. or by the use of antiseptics, is the absence of spores in the filaments which are submitted to the prolonged action of air, of heat, or of diverse chemical agents. The spore is the form of resistance of Bacteridia; it is shielded from the action of most environments, and preserves the properties of the filament which gave it birth. Notwithstanding this resistance to external agents, the germs of the Bacteridia may be modified and attenuated in its virulence in a like manner as the filament. Some spores of *Bacillus anthracis* were kept in contact with sulphuric acid (2 : 100) at 35° C., in a closed tube, and frequent shaken in order to insure contact of the acid and the spores. Every two days a small quantity was placed in a slightly alkaline broth. The cultures thus obtained during the first days killed rabbits and guinea-pigs; those taken at eight and ten days killed guinea-pigs, but were not fatal to rabbits; those taken at fourteen days were not fatal to guinea-pigs. In successive cultures the germs retained their attenuated virulence.

Koch, Gaffky and Loeffler's Method for attenuation. They used the thermostat of D'Arsonval (made by Wiessenegg, in Paris, 64 rue Gay Lussac), which allows only a tenth of a degree variation in temperature. As a culture vessel they used an Erlenmeyer's flask, each flask containing 20 cc. of chicken broth neutralized with carbonate of soda. In each flask they placed a small quantity of virulent anthrax substance, using all precautions against foreign contamination. The flasks were then placed in the thermostat at 42°–43° C., and the degree of attenuation ascertained from time to time by small quantities being taken out and inoculations made upon mice, guinea-pigs, rabbits, and sheep. The authors lay great stress on the use of a pure culture for the test inoculations, "contrary to Pasteur, whose vaccins are often shown to be impure with other Bacteria." Each time a test of the attenuation reached by the cultures was made, new cultures of the same were made in two new flasks of sterilized chicken broth, so that in at least one they should have a pure culture. To fix the degree of attenuation, the flasks were on the fifth day placed at 37° C., where they formed resting spores. The conclusion arrived at was that the protective inoculation methods practised up to the present time are of doubtful advantage.

Staining. The *Bacilli* of anthrax may be stained with the ordinary aniline dyes, and according to general directions. They are, however, difficult to recognize in the tissues, if these are not well decolorized; this is best done by Gram's general method, *q. v.*

Weigert's Double-Staining Method (gentian-violet and picro-carmine). Place sections 2–5 minutes in a one per cent aqueous solution of gentian-violet. Wash in alcohol. Allow sections to float a moment in water, and then place in picro-carmine until intensely stained ($\frac{1}{2}$–1 hour). Wash in alcohol until color has faded somewhat. Oil of cloves, Canada balsam.

Feltz's Method of ascertaining the rôle of earth-worms in the propagation of charbon was to mix some sterilized earth with a fluid containing anthrax *Bacilli*, and to place this in flower-pots. After three weeks the worms were removed one at a time, cut up, after being thoroughly washed, and injected into guinea-pigs.

LITERATURE OF ANTHRAX.

ARLOING (S.): "Influence de la lumière sur la végétation et les propriétés pathogènes du *Bacillus anthracis.*" Compt. Rend. Feb. 9, 1885, p. 378.

ARLOING, CORNEVIN, ET THOMAS: I. "Sur l'inoculabilité du charbon symptomatique et les caractères qui le différencient du sang du rate." Compt. Rend. T. 90, p. 1302 (1883). II. "De l'inoculation du charbon symptomatique par l'injection intra-veineuse, et de l'immunité conférée au veau, au mouton, et à la chèvre par ce procédé." Compt. Rend. T. 91, p. 934.

ARCHANGELSKI: "Ueber Milzbrand." Ctblt. f. d. med. Wiss. 1882, No. 15.

ALADER (v. Rozahegva): "Versuche mit der Pasteur'schen Schutzimpfung gegen Milzbrand in Ungarn." Deutsche med. Wochenschr., 8 Jahrg. 1882, p. 24.

BERT (P.): (Anthrax). "Compt. Rend. Soc. Biol." for 1877. (1879) pp. 19, 20, p. 317, p. 442, p. 465.

BOLLINGER: "Ueber Milzbrand." Ctblt. f. d. med. Wissensch., 1872.

BRAUELL: Ueber Milzbrand." Virchow's Archiv., XI., XIV., XXXVI.

BUCHNER UND ROBERTS: "Ueber d. experim. Erzeugung d. Milzbrand contagiums, aus d. Heupilzen." Archiv. f. exp. Pathol. u. Pharm., Bd. XIII., H. 2, p. 170 (1880).

BUCHNER (H.): I. "Ueber Entstehung des Milzbrandes durch Einathmung." Sitzungsbr. d. königl. bayr. Acad. d. Wiss., 1880. Hft. 3 (Mixed Anthrax spores with dust and allowed white mice to breathe it; in 24 cases they died in from 1–3 days). II. "Ueber die experimentelle Erzeugung des Milzbrand contagiums," *ibid.* 1882, Hft. II.

BOULEY: I. "De la vaccination charbonneuse." Compt. Rend. T. 42, p. 1383; T. 43, p. 190. II.

" Sur l'identité du charbon dans toutes les espèces d'animaux domestique." Compt. Rend. T. 84, p. 993 and 1877.

BURDON-SANDERSON, DUGUN, GREENFIELD, and BENHAM: "Investigations of Anthrax." Journ. Roy. Agric. Soc., 1880, No. 31.

CHAMBERLAND ET ROUX: I., "Sur la vaccinatiere charbon." Compt. Rend. T. 42, p. 1378. II. "Sur les germes charbonneux, *ibid.* T. 92, p. 209. III. "De l'atténuation des virus et leur retour à la virulence," *ibid.* T. 92, p. 429. IV. "La vaccination du charbon, *ibid.* T. 92, p. 662-666. V. "Sur l'atténuation de la virulence de la bactéridie charbonneuse, sous l'influence des substances antiseptiques," *ibid.* T. 96, p. 1088 (1883). VI. "Sur l'atténuation de la bactéridie charbonneuse et de ses germes sous l'influence des substances antiseptiques," *ibid.* T. 96, p. 1410.

CHAMBRELENT ET MOUSSONS: " Expériences sur le passage des bactéridies charbonneuses dans de lait des animaux atteints du charbon." Compt. Rend. T. 97 (1883), p. 1142-1145.

CHAVEAU (A.): I. "Nouvelles expériences sur la résistance des moutons algériens au sang de rate." Compt. Rend. T. 90, p. 1396. II. "Des causes qui peuvent faire varier les résultats de l'inoculation charbonneuse sur les moutons algériens. Influence de la quantité des agents infectants, application à la théorie de l'immunité," *ibid.* T. 90, p. 1526, year 1880. III. "Du renforcement de l'immunité des moutons algériens, a l'égard du sang de rate, par les inoculations preventive. Influence de l'inoculation de la mère sur la réceptivité du fœtus," *ibid.* T. 91, p. 148. IV. "Sur la résistance des animaux de l'espèce bovine au sang du rate et sur la préservation de ces animaux par les inoculations préventives," *ibid.* T. 91, p. 648. V. "Etude experimentale de l'action exercée sur l'agent infectieux, par l'organisme des moutons ou moins réfractaires au sang de rate ; se qu'il advient des microbes spécifiques, introduits directement dans le torrent circulatoire par transfusions massive de sang charbonneux," *ibid.* T. 91, p. 680. VI. "De l'atténuation directe et rapide des cultures virulent par l'action de la chaleur," *ibid.* T. 96, p. 553 (1883). VII. "De la faculté prolifique des agents virulent atténués par la chaleur, et de la transmission par génération de l'influence atténuante d'un premier chauffage," *ibid.* T. 96, p. 612. VIII. "Du rôle de l'oxygène de l'air dans l'atténuation quasi instantanée des cultures virulent par l'action de la chaleur," *ibid.* T. 96, p. 678. IX. "Du rôle respectif de l'oxygène et de la chaleur dans l'atténuation du virus charbonneaux par la méthode de M. Pasteur. Théorie générale de l'atténuation par l'application de ces deux agents au microbes aérobies," *ibid.* T. 96, p. 1471. X. "De la préparation en grandes masses des cultures attenuées par le chauffage rapide pour l'inoculation preventive du sang de rate, *ibid.* T. 98, 1884, p. 73. XI. " Du chauffage des grandes cultures de bacilles de sang de rate, *ibid.* p. 126. XII. "De l'attenuation des cultures virulent par l'oxygène comprimé," *ibid.* p. 1232.

DAVAINE: " Observations sur la maladie charbonneuse." Compt. Rend. (1877) T. 84, p. 1322. Comp. also T. 57 and 59.[1]

DUPLESSIS: " Milzbrandimpfung mit der Pasteur'schen Culturlymph." Journ. d'Agricult. practique. Bd. II., S. 32.

FELTZ: " De la durée de l'immunité vaccinale anticharbonneuse chez le lapin." Compt. Rend. T. 98, Aug. 4, 1884, p. 246, prelim. com. Nov. 6, 1882. II. " Sur la rôle des vers de terre dans la propogation du charbon et sur l'atténuation du virus charbonneux." Compt. Rend. (1882), T. 95, No. 19, p. 859–862.

FOKKER (A. P.): " Zur Bacteriensfrage." Virchow's Archiv. Bd. 88 (1882). II. " Die Identität von *Bacillus anthracis* u. *Bacillus subtilis.*" Ctbl. f. d. med. Wiss. 1880. No. 44.

FRISCH: I. " Die Milzbrandbacterien u. ihre Vegetationen in d. lebenden Hornhaut." Akad.

---

[1] Davaine was the first (1850) to describe *Bacillus anthracis* in " Sang de rate." " Ill y avait en outre dans le sang de petits corps filiform, ayant environ le double en longuer des globules sanguin, ces petits corps n'offraient point de mouvement spontané."

d. Wiss. 74. Bd. III. (1876); II. "Ueber das Verhalten der Milzbrand bacillen gegen niedere Temperatur." Stricker's med. Jahrb., 1879, p. 513.

FRIEDLANDER: "Die Identität v. *Bacillus subtilis* u. *B. anthracis.*" Ctblt. f. d. med. Wiss. 1880.

GOTTI (A.): "Sopra alcuni experimenti di inoculazione carbonchiosa preservativa nei bovini." Memorie della R. Accademia delle Scienze dell' Institute di Bologna. Ser. IV., T. V., p. 734.

GREENFIELD (W. S.): "On *Bacillus anthracis.*" Proc. Roy. Soc., London, 1879–80.

KOCH, GAFFKY, U. LOEFFLER: "Experimentelle Studien über die künstliche Abschwachung der Milzbrand-bacillen und Milzbrand-infection durch Futterung." Mittheil. aus d. kais. Gesundheitsamt. Bd. 2, 1884.

KOCH (R.): I. "Ueber die Milzbrandimpfung. Eine Entgegung auf den von Pasteur in Genf. gehaltenen Vortrag." Kassel u. Berlin, 1882. II. "Zur Aetiologie d. Milzbrandes." Mittl. a. d. kais. Gesundheitsamt, 1881. III. "Ueber Milzbrand und Milzbrandimpfung." 54 Vers. deutsch. Naturf. u. Aertze. Salzburg, 1881. IV. "Milzbrandversuche." Mittheilung a. d. kais. Gesundheitsamt, 1881, Berlin.

KOUBASSOFF: "Passage des microbes de la mère au foetus." Compt. Rend. T. 100 (1885), p. 372.

Lutz: "Eine Milzbrand epidemie bei Menschen.'" Aerztl. Int. Bl., 1881, 21.

Osol (K.): "Das Anthrax Virus." Ctbl. f. d. med. Wiss. 1884, p. 401–404.

Pasteur et Joubert: "Etude sur la maladie charbonneuse." Compt. Rend. T. 84, p. 900 (1877). II. "Charbon et septicémie." Compt. Rend. T. 85, p. 101; also p. 61.

Pasteur (L.): I. "Sur l'aetiologie du charbon." *Ibid.* T. 91, p. 86 (1880). II. "Experiences tendant à démontrer que les poules vaccinées pour la choléra sont refractaires au charbon." *Ibid.* T. 91, p. 315 (1880). III. "Sur l'etiologie des affection charbonneuses." *Ibid.* T. 91, p. 455 (1880). IV. "Sur la non récidive de l'affection charbonneuse." *Ibid.* T. 91, p. 531 (1880). V. "Nouvelles observations sur l'etiologie et la prophylaxie du charbon." *Ibid.* T. 91, p. 697 (1880). VI. "Announcement that Anthrax germs were brought to surface of ground by earth-worms." Trans. d. Acad. d. méd. à Paris, 17 Mai, 1881. VII. "La vaccination charbonneuse. Réponse au Dr. Koch." Revue scientifique, Jan. 20, 1883.

Pasteur, Chamberlain, et Roux: "De l'atténuation des virus et de leur retour à la virulence." Compt. Rend. 28 Feb. 1881. II. "Le vaccin du charbon." *Ibid.* T. 92 (1881), p. 666. III. "Sur la longue durée de la vie des germes charbonneux et sur leur conservation dans les terres cultivées." *Ibid.* Jan. 31, 1881.

PERRONCITO: "Ueber die Tenacität des Milzbrand Virus in seinen beiden Gestalten, als Spore u. Bacillus anthracis." Revue fur Thierheilkunde u. Thierzucht, 1883, No. 11.

POINCARÉ: "Sur la production du charbon par les pâturages." Compt. Rend. T. 91, p. 179 (1880).

POLLENDER: "Microscopische und chemische Untersuchungen d. Milzbrandblutes." Caspers Vierteljahrsschrift f. gerichtl. Medecin. 13.

PRAZMOWSKY (A.): "Ueber den genetischen Zusammenhang der Milzbrand und Heubakterien." Biol. Centralblatt. 1884, No. 13.

RODET (A.): "Sur la rapidité de la propagation de la Bactéridie charbonneuse inoculée." Compt. Rend. 10 Avril, 1882. Journ. de Micrographie, 1882, p. 408.

ROLOFF: I. "Milzbrand Entstehung u. Bekampfung." 1882. II. "Ueber die Milzbrandimpfung und entwickeln d. Milzbrandbacterien." Arch. f. wiss. u. pr. Thierheilkunde. Bd. 9, Hft. 6. p. 459.

ROSSINGOL (H.): "Les nouvelles experiences de Pouilly-le-Fort. L'immunité conférée par la vaccination, practiquée avec le virus charbonneuse attenué de M. Pasteur, est-elle transmissible de la mère au foetus?" Angers (Lechese et Dolbeau), 1883.

STRAUSS ET CHAMBERLAND: "Passage de la bacteridie charbonneuse de la mère au foetus." Compt. Rend. T. 95, No. 25 (1882).

SZPILMAN (J.): "Ueber das Verhalten der Milzbrand bacillen in Gasen." Hoppe-Seyler's Zeitschrift f. physiol. Chemie. Bd. IV., p. 350 (1880).

TOUSSAINT (H.): I. "De l'immunité pour le charbon, aquise à la suite d'inoculation préventives." Compt. Rend. T. 91, p. 135 (1880). II. "De l'immunité pour la charbon, aquise à la suite des inoculations prêventives. Procédé pour la vaccination du mouton et du jeune chien." *Ibid.* T. 91, p. 303 (1880).

WEIGERT: (Staining method). Virchow's Archiv. Bd. 81.

CHOLERA.

(Comma Bacillus. Koch.)

The importance of a knowledge of the Bacteria inhabiting the healthy body, as well as a proper preliminary training in the technique of Bacteria investigation, is strikingly shown in connection with the microbe of cholera. Lewis claimed that a *Bacillus* resembling that of cholera could be found in the mouth; in regard to which Dr. Koch shows, in a very few words, that the form referred to has been known for several years, that it differs from the *Comma Bacillus* in being longer, more slender, and not so blunt at the ends, and further differs in the important particular that it does not form the characteristic colonies in weakly alkaline peptone gelatine. Again, Finkler and Prior claimed to have found the *Comma Bacillus* in the stools of *cholera nostras* patients; Koch obtained some of their culture material, and found in it four different microbes, of which one resembled slightly the *Comma Bacillus*, but was larger and plumper, and in its mode of growth quite different, growing much more rapidly in gelatine or on potato, and showing unmistakable differences in the form assumed in the cultures.

Koch's Method for diagnosing the cholera *Bacillus*.[1] Microscopical study of the intestine of cholera patients showed, especially where the Peyer's patches were reddened at the edges, an inwandering of Bacteria; they were found partly in the glands, partly between the epithelium and the basement membrane, raising up the epithelium. Some of the Bacteria had a peculiar appearance as to size and shape, by which they could be distinguished. The contents of the intestine exhibited a great variety of Bacteria. He selected two acute, uncomplicated cases, in which no blood had yet appeared in the stools, and compared the *Bacilli* of the intestinal contents until he found

---

[1] Taken from Koch's report at the Cholera Conference, July 26, 1884.

some similar to those in the glands. These he named *Comma Bacilli*. They are smaller than the *Bacilli* of tuberculosis, about two thirds as long, but much plumper and thicker, and with a slightly bent appearance, not usually any greater than that of a comma; sometimes the curve is doubled, so that an S shape results. This arises from the two individuals formed by fission adhering together. In pure cultures another very characteristic developmental form is seen. It consists of more or less long, screw-shaped filaments, not straight, not wavy, but having a great resemblance to the *Spirochæte* of recurrent fever. Koch thinks that if a person had both forms together upon a slide he would be unable to distinguish between them. From this he concludes *Comma Bacilli* to be an intermediate form between *Bacillus* and *Spirochæte*. The *Comma Bacillus* behaves like a piece of *Spirillum*, or like the short specimens of *Spirillum undula* which do not make an entire screw, but only short rods more or less bent. Cultivated in meat broths, the cholera *Bacilli* increase with enormous rapidity and in great abundance, and are unusually lively. They grow very luxuriantly and rapidly also in milk, but do not cause it to curdle, and do not precipitate the casein, as do many other Bacteria. The milk appears unchanged, but if one takes a small drop from the surface and examines it microscopically, it is found to teem with *Comma Bacilli*. They grow also very vigorously in blood serum. Another good culture medium is peptone gelatine. In the latter they assume an entirely characteristic and peculiar form of colony formation, such as Koch had never seen elsewhere. The colony appears when very young as an exceedingly pale, small drop, not perfectly round in outline (as is usually the case with Bacteria colonies in gelatine), but having a more or less irregular border, presenting a bored-out, stellate, or ragged, toothed edge. They have also a somewhat granular aspect, which is not of so regular a character as in other Bacteria colonies. As the colony grows, the granulations become more apparent, seeming at length like a little heap of glass particles. Upon further growth the gelatine is dissolved in the immediate vicinity of the colony, which sinks deeper into the mass. There is thus formed a funnel-like pit, in the middle of which the colony may be seen as a small, white speck. This behavior is also peculiar to *Comma Bacilli*, at least it is only seen in very few other forms of Bacteria, and nowhere so outspoken. One can best see the sinking of the colony if he prepares a pure culture as follows: Search out upon the gelatine plate, with a weak power objective, a suitable colony, pick it up with a sterilized platinum wire, and inoculate a properly prepared gelatine culture in a test tube plugged with cotton.

As soon as the culture begins to develop, one sees a small, funnel-shaped pit being formed at the point of inoculation. Soon the small colony itself becomes apparent, but remains above a deep, sunken spot, which appears to be partially fluid gelatine; there is also the appearance as if an air bubble hung over the colony. It would seem that the vegetation of the *Bacilli* not only dissolves the gelatine, but rapidly thins the fluid formed. There are numbers of other Bacteria which dissolve the gelatine in cultures, but in none other is there found such a deepening, nor such a bladder-like cavity on the surface. The solution of the gelatine gradually progresses, and in about a week the entire contents of the tube is dissolved. Special importance is placed upon these peculiarities, as they serve to distinguish the *Comma Bacilli* from other *Bacilli*. *Comma Bacilli* may also be cultivated in cultures made with Agar-Agar, which they do not dissolve as they do gelatine. Cultures should also be made upon boiled potato, when they form colonies resembling those of the *Bacilli* of glanders. The latter form upon the potato a thin, pulpy, brown coating. The colony of the cholera *Bacillus* differs from this in not being so intensely brown, but rather more of a clear, grayish-brown. The *Comma Bacilli* thrive best at a temperature between 30° and 40° C., but they are not very susceptible to lower temperatures; thus they will develop very well, although somewhat slowly, at 17° C. Below 17° C. the development is very slight, and seems to cease at 16° C. In this respect the cholera *Bacilli* correspond completely with the *Bacilli* of anthrax, which also have this temperature as a limit to their growth. Koch made experiments with still lower temperatures to ascertain whether these would not only stop the development, but kill the microbes. For this purpose a culture was exposed for an hour to a temperature of — 10° C.; the *Comma Bacilli* were here completely frozen for this length of time. An addition made from this frozen culture to fresh gelatine showed not the slightest change in development; they stood the freezing well. They, however, ceased to grow after having the air shut off from them. This can be shown in a simple manner. After pouring the gelatine culture containing the microbes upon a glass slide, and just before it stiffens, place upon its surface a sheet of mica, which has been split to the utmost thinness possible, and which covers at least one third of the gelatine surface in the middle. The mica sheet, by its elasticity, allows itself to come into complete contact with the gelatine, and shuts out the air from the covered spot. It is now seen, as the colonies begin to develop, that growth only occurs where the gelatine is uncovered, extending under the edge of the mica only so far as the air is admitted, but under the mica itself the microbes do not grow at all. They here remain as unusually small colonies, invisible to the naked eye, which have grown this much from using up the oxygen contained in the gelatine, and ceasing to grow when this is consumed. This investigation may also be made in another way. A glass containing a gelatine culture medium, which has been inoculated with *Comma Bacilli*, is placed under the air pump, which is then exhausted. Other glasses prepared in the same way and at the same time are allowed to remain outside. Those under the pump will not grow. If, later, these be restored to the air, they begin at once to develop. They have not been killed, but appear able to increase only under the influence of oxygen. The effect is similar if cultures are placed in an atmosphere of carbonic acid, while the control cultures in the air grow as usual, those in the gas remain entirely undeveloped. They are not destroyed by this treatment, but begin to grow if restored to the air after having been for a long time in carbonic acid. Under favorable circumstances the *Comma Bacilli* grow very rapidly. This vegetation quickly reaches a maximum, and there remains stationary for a short time, and then rapidly declines. After dying, the *Comma Bacilli* lose their form, appearing swollen or shrunken, and refusing to take staining altogether, or only slightly. The peculiar vegetation behavior is best observed by placing upon damp earth, substances which, while rich in *Comma Bacilli*, contain at the same time other Bacteria, *e. g.*, the intestinal contents or cholera dejections are placed upon damp earth or linen, and kept in a damp place. The cholera microbes here increase in a most astonishing way for a short time. At first the other Bacteria seem to be stifled by the *Comma Bacilli*, which build for themselves a natural pure culture, and one can find, by examining microscopically the surface of the damp earth, or of the linen, preparations which show almost none of the common Bacteria. They do not retain this luxuriant growth long; after two or three days they begin to die, and the other Bacteria begin to increase. This behavior is exactly similar to that in the intestine, where they also have a rapid increase; but if this peculiar vegetation process, which only lasts for a short time, is over, and especially if there has been a transudation of blood into the intestine, the *Comma Bacilli* decrease in numbers, and the filth Bacteria begin to develop more rapidly. From this, Koch infers that if the cholera *Bacilli* are brought from the intestine into a putrid fluid which contains much of the products of decomposition of other Bacteria, and especially of filth Bacteria, they will not develop well, but will soon die. Sufficient investigation has not been as yet made upon this point, which is important in so far that it is well to know whether the *Comma Bacilli* find a good culture medium or a very poor one when they come into a water-closet or sink. In the first case they would increase and require to be destroyed by some disinfectant; in the latter case they would die, and no further disinfectant be needed. Dr. Koch accepts the latter conclusion.

The *Comma Bacilli* thrive best in fluids which do not contain too little nourishment; thus, when a meat-broth culture was thinned fivefold, it was found to be no longer useful as a culture fluid. In other experiments they still continued to grow when the fluid was thinned tenfold. These experiments must evidently be repeated in a more extensive and systematic manner, but in any case it is known from the results obtained that one dare not go too far in the dilutions, and that the *Comma Bacilli* demand a certain concentration in their nourishing medium. In regard to cultures it is further to be remarked that the culture gelatine and the meat broths must not be of an acid reaction. As soon as the slightest trace of acidity shows itself, the development of the *Comma Bacilli* is greatly impeded. If the reaction becomes evidently acid, then the development ceases. This is, however, not true of all acids; thus, the cut surface of a boiled potato has an acid reaction from the presence of malic acid. Here, however, the cholera microbes grow luxuriantly.

There are, at any rate, a number of acids which do hinder the growth. This is apparent in meat broth if lactic acid or an acid phosphate be added.

Nicati and Rietsch's Method. — Pure cultures of *Comma Bacilli*, according to these investigators, present a characteristic odor which is not putrid or disagreeable, but somewhat ethereal ("éthéré"). By means of a Pasteur's filter, they removed from an eight-days-old pure culture in broth or nutritive gelatine (Koch's formula) all the *Bacilli*, and injected the filtrate into the blood-current of a dog (jugular and crural vein), which had, as a consequence, all the symptoms of cholera.

Bochefontaine's Method. — I give this method, not because I think others will care to or should repeat it, but as an example of the heroism with which investigators carry on researches for the public good, like impetuous soldiers, often taking unnecessary risks, but requiring a higher grade of fortitude from the lack of all excitement, and in full knowledge of the possible results.

Bochefontaine made five large, soft pills by incorporating the serous alvine dejections of a cholera patient with gum and lycopodium. These pills he took successively, drinking afterwards a large glass of water. Three hours later the skin became hot; pulse 100 (66 being in his case normal); the fever persisted for twenty-four hours, the pulse going up at times to 120; slight nausea;

insomnia for three hours; disurea part of the time; slight convulsions of the muscles of the legs, of the forehead, and of the fingers of the right hand; loss of appetite; and constipation for twenty-four hours. He then took a glass of purgative alkaline water, and was all right again. The dejections used contained microbes of all kinds, and among them the *Comma Bacillus*. This experiment shows that the injection of the liquid diarrhœa of cholera, containing *Comma Bacilli*, into the stomach does not necessarily produce cholera.

LITERATURE.

BIEDERT (H.): Ueber den Cholera-bacillus in popular wissenschaftlicher Weise. Kölnischen Zeitung, No. 314, 1884.

BIEDERT (PH.): " Die Reinkulturen in Reichs-Gesundheitsamt und der *Cholera Bacillus*." Berlin, 1885, 8vo.

BIENSTOCK: (Cholera Bacillen). Zeitschrift f. klin. Med. VIII. (1884), Hft. 1, 2.

BOCHEFONTAINE: " Expérience pour servir à l'étude des phenomènes déterminés chez l'homme par l'ingestion stomacale du liquid diarrhéique du cholera." Compt. Rend. T. 99, Nov. 17, 1884, p. 845.

BÖRNER (P.): I. Die bisherige Thätigkeit der Herren Finkler und Prior zur Aetiologie der Cholera, und ihre Entgegung wider R. Koch — in der Kölnische Zeitung." Deutsch. med. Wochenschr., No. 47, 1884, p. 770. II. Weitere Beiträge zu der Discussion über die Cholera Bakterien." *Ibid.* No. 48, p. 788, Nov. 27, 1884.

BUCHNER (H.): "Ueber die Koch'schen u. Finkler-Prior'schen Komma-bacillen." Sitzungsbericht der. Ges. f. Morphologie u. Physiol. in Munchen. Sitzung. v. 13, Jan., 1885. II. "Bemerkungen zu Flügge's Kritik der Emmerich'schen Cholera-untersuchungen." Sep. Abdruck aus d. Berliner klin. Wochenschr. 1885. No. 15.

EMMERICH (R.): "Ueber die in choleraleichen und cholerakranken gefundenen Pilze." Deutsch. med. Wochenschr. 1884. No. 50; also Archiv. f. Hygiene, Jan., 1885. The Lancet, Dec. 27, 1884.

ERMENGEN (E. VAN): I. "Contribution a l'étude du Microbe du Cholera Asiatique. Recherches sur un micro-organisme, découvert par MM. Finckler et Prior dans le cholera sporadique," Journal de Micrographie, No. XI. (Nov., 1884), p. 595. II. "Recherches sur le bacille virgule du choléra asiatique, conclusions principales du travail présenté a la Société Belge de Microscopie dans sa séance de 26 Octobre, 1884. Ext. in Deutsch. Med. Wochenschr., No. 46, p. 749. Published separately. Bruxelles, 1884. pp. 37. 8vo, avec 2 plch. photogr.

FINKLER U. PRIOR: "Untersuchungen über Cholera nostras." Congress des Naturalists Allemand, a Magdebourg, Séance du 20 Sept., 1884. Deutsche. med. Wochenschr., No. 36 (1884), p. 581: also, No. 59, p. 632–634. Kölnischen Zeitung, Nov. 11, 1884. Tageblatt der Naturforchersers, 1884; p. 218–223.

KLAMANN: "Bacilli d. Cholera nostras." Tageblatt der Naturforscherers, 1884, p. 223.

KLEBS (E.): "Ueber Cholera Asiatica" (Cholera bacillen), Basel. 1885. (B. Schwalbe), pp. 18, gr. 8vo, mit Holzschnitten. cf. Schweizer ärztl. Corresp. Bl. 1884. No. 23.

KLEIN (E.): On Cholera Bacilli. Journ. of Science, VI., 1884, p. 510.

KOCH (R.): I. Bericht der deutschen wissenschaftlichen Commission zur Erforschung der Cholera, March 4, 1884. Abstract in Fortschritte der Med. 1884, Bd. 2, p. 68. II. Bericht d. Conferenz zur Erörterung der Cholerafrage. Berl. klin. Wochenschr. 1884. No. 31, ff. July 26, 1884. Abstract in P. Börner's Reichs-Med. Kalender f. Deutschld. Theil. II. 1885. "Ueber die Cholera bacterien." Deutsche Med. Wochenschr., No. 32 u. 32 a, 1884; also No. 45 (Nov. 6), p. 725.

LANKESTER (E. RAY): "On Comma bacilli." Nature, 1885, XXXI., p. 163–171, 6 Figs. cf. Pall Mall Gazette, Oct. 6, 1884, pp 1, 2. Times, Nov. 19, 1884.

Lewis (T. R.) : "Memorandum on the 'Comma-shaped Bacillus' alleged to be the cause of cholera." Med. Times and Gazette, Sept. 20, 1884. The Lancet, Sept. 20, 1884, p. 513.

Marey : "Les eaux contaminées et le Cholera." Compt. Rend. T. 99, Oct. 27, 1884, p. 667.

Nicati und Reitsch : I. "La vitalité du Microbe de Cholera." Revue Scientifique, No. 21, 1884. II. "Odeur et effets toxique des produits de la fermentation produite par le bacilles en virgule." Compt. Rend. T. 99, 1884. Semaine méd. 1884. No. 38. Deneke deutsche med. Wochenschr. 1885. No. 3. Chemisches Centralbl. 1885. No. 5.

Nicati (W.) : "Cholera et Cholemie." Compt. Rend. T. 99, Nov. 24, 1884, p. 929.

Pacini (F.) : "Nuove Osservazioni microscopische sul Colèra." Memorie inedite racotta e publicate per cura di A. Bianchi. Milano, 1884.

Pelletan (J.) : "Le 'Kommabacillus' Koch." Journal de Micrographie. T. 8, p. 475.

Pettenkoffer (Max v.) : "Zur Aetiologie oder Infectionkrankheiten. 1881, pp. 333–352. cf. Bot. Centralbl. IX., 1882, p. 25. (On the parasitic nature of cholera.)

Pouchet (G.) : "Sur la presence des sels biliaires dans le sang des cholerique et sur l'existence d'un alcaloide toxique dans les dejections." Compt. Rend. T. 98, Nov. 17, 1884, p. 847.

STERNBERG (G. M.) : " The Comma Bacillus of Koch." Science, Feb. 6, 1885, p. 109.

TAXIS ET CHAREYRE : "La Bacille du Choléra." Journal de Micrographie, T. VIII., p. 397, 444.

UNGAR : " Ueber das Verhalten des von Finkler und Prior in den Stuhlentlechungen bei der Cholera nostras gefundenen Bacillen." Kölnischen Zeitung, No. 323, 1884.

VENTUROLI (M.) : "Il bacillo-virgola di Koch e la microscopia." Bologna (Tip. Arcvvscovile) 1884, 16 pp. 16mo.

WATERS (G.) : " The Comma-shaped Bacillus a zymotogenic, not a pathogenic, entity." Med. Times and Gazette, 1884, Nov. 8.

GLANDERS.

Bacillus malandriæ (Israel).

Loeffler and Schutz Method. — A firm culture medium is made from horse-blood serum, and inoculated with the microbes of a fresh glander knot. After three days the surface of the serum shows numerous small, transparent drops; in these are found fine *Bacilli*, such as are found in the diseased lung, liver,.spleen, and nasal wall. For staining, use a concentrated aqueous solution of methyl-blue; treat with highly diluted acetic acid, then dehydrate with absolute alcohol, cedar oil, balsam. Inoculation experiments need further repetition. In this disease the *Bacilli* are very difficult to find. The investigation must be carried on with an oil-immersion objective, and an Abbe's condenser. There is no peculiar staining method by which we can find the *Bacilli* of glanders in living animals, since it is impossible to stain them without, at the same time, staining the Bacteria normally to be found in the nasal mucus.

LITERATURE.

BOUCHARD, CAPITAN ET CHARVIN : Revue medicale française, Dec. 30, 1882.

GALTIER (V.) : "Inoculation de la morve au lapin ; destruction de l'activité virulente de la morve par la déssication, transmission de la morve par l'inoculation de la saline." Compt. Rend. T. 91, p. 475. II. "Inoculation de la morve au chien." *Ibid.* T. 92, p. 303.

ISRAEL (O.) : "Ueber die Bacillen der Rotzkrankheit" (Vortrag, gehalten in der Sitzung der Gesellschaft der Charitié Aerzte 1, Feb. 1883). Berliner klin. Wochenschrift, 1883, No. 11, p. 155.

LOEFFLER AND SCHÜTZ : "Ueber den Bacillus des Rotzes." Mittheilung aus dem deutschen Reichsgesundheitsamt. Deutsche med. Wochenschrift, No. 51, 1882. Berliner klin. Wochenschrift, Jan. 1883, p. 27.

STRUCK : Deutsche med. Wochenschr. Nos. 51, 52, Dec., 1883.

Vulpian et Bouley: "Sur une note communiquée a l'Academie sur la culture du microbe de la morve et sur la transmission de la maladie à l'aide des liquides de culture, par MM. Bouchard Capitan et Charrin." Bull. de l'Acad de Medicine, 1883. N. 41, Seance du 30 Oct.

"The Bacillus of Glanders, New Mounting Medium." The Microscope. Vol. IV., 1884, No. 4, p. 79.

Wassilieff: "Die Bacillen des Rotzes und ihre Bedeutung für die Diagnose." Deutsche med. Wochenschrift, 1883, No. 11.

HOG CHOLERA.

"Pneumoenteritis suis (bacillaris)." *Bacillus minimus* (Klein). — "Swine plague Schizophyte." Detmers.

As in so many other cases there is a difference of opinion as to whether the pathogenic microbe of hog cholera is a *Bacillus*, or a dumb-bell *Micrococcus* (*Diplococcus*). Klein still maintains the correctness of his conclusions, and claims that Pasteur is in error in all his experiments which point to its being a *Micrococcus*. Klein takes no notice whatsoever of the investigations of Detmer's, of which Pasteur speaks as follows: —

"La vérité historique toute fois m'oblige à déclarer qu 'en 1882, et également au mois de mars, le microbe du rouget avait été signalé à Chicago, en Amérique, par le professeur Detmers, dans un travail qui fait grand honneur à son auteur." Compt. Rend. T. 97, p. 1164.

LITERATURE.

DETMERS (H.): "Remarks on a pathogenic Schizophyte." Micrococci of swine plague or hog cholera. Ann. and Mag. of Nat. Hist. 5th ser. vol. VII., p. 471. Extract from 'Science,' May 7, 1881, read before State Microscopical Society of Illinois, April 8, 1881. — Amer. Naturalist, March, 1882. — Journal de Micrographie, 1882, vol. VI., pp. 172, 223, 496.

EGGELING : "Ueber den Rothlauf der Schweine." Fortschritte der Medicin, Bd. 1, Ar. 23, p. 793, 1883.

KLEIN : "Report on Infectious Pneumoenteritis of the Pig." Report of the Med. Office of the Privy Council, 1877–78 : London. "Die Bacterien der Schweinseuche." Virchow's Archiv, Bd. 79, p. 468.

PASTEUR : (Microbe d. Schweinrocken) Mittheilungen d. Akad. d. Med., Nov. 27, 1883.

PASTEUR ET THUILLIER : "La vaccination du rouget des porcs à l'aide du virus mortal atténue de cette maladie." Compt. Rend. T. 97, p. 1163.

SALMON : (On swine plague). Rep't Dep't Agricult. Washington, 1880–81 ; see also article on "Contagious diseases of domestic animals." *Ibid.* for 1884.

HYDROPHOBIA.
Bacillus lyssæ (Pasteur).

There is still room for careful research in regard to the specific microbe of hydrophobia, opinions differing as to whether it is a *Bacillus* or a *Micrococcus*. According to Pasteur it is the former; according to Gibier and Rabe the latter.

Gibier's Method of Attenuation. — The virus is, according to this plan, deprived of a portion of its virulence by exposure to very low temperatures. It was found that no change was produced by cold from 0° to 35° C., but when the virus was kept for eight hours at a temperature of 35° C., the animals did not all die; at 40° to 43° C. the dogs and rabbits resisted its effects, and were only rendered slightly sick. Gibier did not prove whether this inoculation gave an immunity to hydrophobia. His method of inoculation was to make a hole, by means of a small drill, in the median line of the skull; the virus being injected by means of a hypodermic syringe passed through the opening thus made. The dogs are kept quiet by means of a hypodermic injection of morphia at the base of the ear.

Pasteur's Method of Inoculation. — Pasteur trepanned the animal, and injected the virus, mixed with water, upon the brain by means of a Pravaz syringe. After a period of from fifteen to twenty days, the animals died of hydrophobia, and a bit of their brain was always capable of producing hydrophobia in other animals.

M. Pasteur sums up the results of his experiments as follows: —

I. If the virus of rabies is passed from a dog to a monkey, and then from one to other monkeys, it gradually becomes weaker. If it is then injected into a dog, rabbit, or guinea-pig, it remains in the attenuated condition.

II. The virulence of the poison is increased when it is passed from rabbit to rabbit, or from guinea-pig to guinea-pig. If in this "exalted" condition it is passed into a dog, it gives a rabies which is always mortal in effect.

III. Although one can thus increase the virulence of the poison by passing it from one to another rabbit, it is necessary to do several times if one is making use of a virus which has been attenuated by a monkey.

Thanks to these observations, Pasteur has been able to preserve an organism from the effects of more active virus, by the use of that which is less so. Here is an example: — Virus made more powerful by passage through several rabbits is inoculated into a dog, but it is inoculated into the dog at every stage of the experiments on rabbits; the result is, that the dog becomes entirely refractory to the most virulent virus."

"Une des plus grandes difficultés des recherches sur la rage consiste, d'une part, dans l'incertitude du développement du mal à la suite des inoculations ou des morsures, et d'autre part dans la durée de l'incubation, c'est-à-dire dans le temps qui s'écoule entre l'introduction du virus et l'apparition des symptômes rabique. C'est un supplice pour l'expérimentateur d'être condamné à attendre, pendant des mois entiers, le résultat d'une expérience, quand le sujet en comport de très nombreuses. On apprendra donc, je l'espère avec un vif intérêt, que nous sommes arrivés à diminuer considerablement la durée d'incubation de la rage et à la communiquer à coup sûr. On arrive à ce double résultat par l'inoculation directe à la surface du cerveau, en ayant recours à la trépanation et en se servant comme matière inoculante de la substance cérébrale d'un chien enragé, prélevée et inoculée à l'état de pureté. Pasteur, Compt. Rend. T. 92, p. 1260.

Babe's Staining Method, for making cover-glass preparations of saliva, was by the use of methyl-violet, according to general rules. Fuchsin solutions, made alkaline with aniline oil, may also be used.

In what manner the specific microbes are to be distinguished from those normally existing in the saliva, we are not told in any of these methods, and there appears to have been no such isolation.

Some of Koch's remarks in reference to Pasteur's methods would therefore seem to be justified in this instance; for example, the following (taken from Koch's article " Ueber die Milzbrandimpfung : Eine Entgegung auf den von Pasteur in Genf. gehaltenen Vortrag," Berlin 1882, p. 5, 6, 7) : " Pasteur takes, in the first place, the position that all infectious diseases are parasitic, produced by microbes, and he appears to regard the conditions advanced by me for identifying the micro-organism and investigating its presence in the body as unnecessary. For example, Pasteur makes no declaration whether he, in the investigation of the disease called by him *nouvelle maladie de la rage*, examined microscopically, above all things, the sub-lingual gland, for the presence of specific microbes. In this case such an investigation is indispensable, for it is known that in rabies the sub-lingual glands contain the infectious material, and since the tissues of the sub-lingual glands are not ordinarily the seat of Bacteria, the supposed specific microbe may most certainly be found here. Pasteur, however, in conveying the disease from a child, dead of rabies, to an animal, used as a vaccine, not the tissue of the sub-lingual gland, but the saliva itself, in which it is known an immense number of bacterial forms are to be found, pathogenic forms in healthy men. The microbes of hydrophobia, which they hoped to find at that time, and which are still sought for in vain, Pasteur did not find, but instead, Bacteria were found which were considered as the cause of a new disease. Upon examining more closely this 'new disease,' it was soon found to be the already long known septicæmia of rabbits."

The very fact that Pasteur laid himself open to such criticism as the above shows how extremely careful the investigator must be, to leave no method untried before announcing to the world a new discovery connected with pathogenic Bacteria.

LITERATURE.

CHAMBERLAND ET ROUX : " L'organisme microscopique trouvé dans la maladie nouvelle provoque par la saliva d'un enfant mort de la rage." Journ. de Pharm. et Chém. T. 3.

GIBIER (P.): I. "Recherches sur la rage" Compt. Rend. T. 98. 1884. p. 55. II. Recherches experimentales sur la rage. 1°. les ouiseaux contractent la rage. 2°. ils guerissent spontanément *Ibid.* T. 98, p. 531. Feb. 25, 1884. III. "Recherches sur le rage." *Ibid.* T. 96, p. 1701 (1883).

LINDEMANN : "Zur Pathogenese der Lyssa humana." Berl. klin. Wochenschr. 1879. No. 4.

PASTEUR, CHAMBERLAND AND ROUX : " Nouvelle Communication sur la rage." Compt. Rend. T.

98, No. 8, p. 477. (1884) *Ibid.* No. 20, p. 1229. II. "Le virus de la rage atténué." Journal de Micrographie, T. 8, p. 345.

PASTEUR, CHAMBERLAND, ROUX, ET THUILLER: "Nouvelles faits pour servir a la Connaissance de la rage." *Ibid.* 1882. T. 95, No. 24. "Sur la rage." *Ibid.* T. 92 (1881), p. 1259.

PERCHERON (G.): "La rage et les expériences de M. Pasteur." Paris (Firmin Didot), 1884, 149 pp., 18mo.

RABE: Revue für Thierheilkunde. No. 1. Bd. VII.

LEPROSY.

Bacillus lepræ (Hansen).

The staining of this microbe in cover-glass preparations is only possible when done immediately after drying. According to Baumgarten, the staining is possible after a length of time, provided the strength of the solution (fuchsin) be diminished. If too great a time has elapsed, the *Bacilli* will take no staining. The lepra *Bacilli* are the only yet known organisms which possess the peculiarity of tubercle *Bacilli* of not giving up in strongly acid solutions the stain which they have previously taken on. Like the *Bacilli* of typhoid fever, the microbe of leprosy shows small unstained portions. Gentian-violet, methyl-blue, methyl-violet, and fuchsin, stain *Bacillus lepræ;* Bismarck-brown does not.

Babe's Method for Staining. — Prepare according to general directions, and stain in a solution of rosanilin-chlorhydrate in aniline water (F. 36). Decolorize in a nitric acid solution (33⅓ per cent). Stain the nuclei present in the preparation with methyl-blue. A better method, showing the difference of taking and losing the stain of lepra and tubercle *Bacilli*, is, —

Baumgarten's Differentiation Method. — 1. *For Cover-Glass Preparations*, add five drops of a saturated alcoholic solution of fuchsin to a watch-glass of distilled water. Upon this mixture allow freshly prepared cover-glasses to float for six or seven minutes. Decolorize for fifteen seconds in acidulated alcohol. (Nitric acid 1 part, alcohol 10 parts). Restain with an aqueous solution of methyl-blue, wash in water, pass through absolute alcohol, oil of bergamot, and mount in balsam. Study with a one-twelfth oil immersion objective with open condenser. *Tubercle Bacilli do not stain by this method!*

2. *For Sections*, allow the sections to remain twelve or fifteen minutes in above staining fluid, decolorize for thirty seconds in acid solution, wash three or four minutes in water, absolute alcohol, etc., or allow the sections to remain for two or three minutes in fuchsin-aniline water (F. 17.), decolorize for thirty seconds in above acid mixture. Restain with a concentrated aqueous solu-

tion, of methyl-blue for two or three minutes; absolute alcohol, oil of bergamot, Canada balsam. (*Tubercle Bacilli do not stain by this method!*)

3. "Cover-glass" and dry preparations are to be made according to Baumgarten's method for tubercle *Bacilli* (*q. v.*) (although the above somewhat dilute staining fluid may be used). Lepra *Bacilli* assume, after two or three minutes, a red color, while those of tuberculosis appear entirely colorless.

Neisser's Methods. — Neisser remarks, in regard to staining (l. c. p. 524). In preparations hardened in alcohol and not stained, the *Bacilli* are invisible. Strong acetic acid renders them somewhat visible. Caustic potash (1 : 12) gives better results. They are well stained in gentian-violet or methyl-violet, and better in fuchsin. At best, they are difficult to stain, and the best results are to be obtained in specimens handled in weak caustic potash solution, then stained and washed in acidulated alcohol. The *Bacilli* in the gentian-violet preparations are to be distinguished by their red tone from the nuclei, which are blue. Dahlia gives in acid solution useful preparations. Methyl-blue is useful only in showing the formation of vacuoles in the old lepra cells. A preparation of eosin and hæmatoxylin (F. 14.) is useful. The sections after staining are to be washed in water and decolorized in alcohol. The nuclei show themselves stained a beautiful blue, the ordinary cell protoplasm eosin-rose, and the protoplasm of cells containing *Bacilli* a light orange, so that the *Bacilli* can be found with weak powers. Some of his preparations lost their color after twenty-four hours, some retained it for two years, differences which he cannot explain.

Neisser's Culture Methods consisted of removing a lepra tubercle which was still covered with sound skin after having thoroughly cleaned the latter, employing in the process of extirpation the greatest precautions against foreign infection. Cultures were then made from this, in closed slides with blood-serum, or in sterilized test tubes with blood serum, or with alkaline meat-extract solution. All preparations were kept in an oven at 35°–39° C.

LITERATURE.

ARNING (E.) : "Ueber das Vorkommen des Bacillus lepræ bei Lepra anæsthetica sive nervosum." Virchow's Archiv, T. 90, p. 170.

BABES (V.) : (Budapest) " Etude comparative des bactéries de la lepra et de la tuberculose." Compt. Rend. de l'acad de Sc. 17 Sept., 1883. II. "Note sur le rapport des bacilles de la tuberculose et de la lèpre avec les surfaces tegmenteuses." Compt. Rend. des Sci. de la Soc. de Biol, 1883, avril 21. III. " Observations sur la topographie des bacilles de la lèpre dans les tissus et sur les bacilles du choléra des poules." Archiv. de phys. normal et path. 1883. No. 5.

BAUMGARTEN : " Ueber Untersuchungsmethoden zur Unterscheidung von Lepra- und. Tuberkel-Bacillen." Zeitschrift für wissenschaftliche Mikroskopie. Von Dr. Behren's. Bd. 1. H. 3. p. 307.

CORNIL (V.) : " Note sur la siege des bactèries dans la lèpre et sur les lesions des organes dans cette maladie." Bull de l'acad. d. Med. No. 9, also in Gaz. Med. d. Paris. No. 44, and Union Med. No. 178–179.

DAMSCH : " Uebertragungsversuche von Lepra auf Thiere." Virchow's Archiv. Bd. 92 Heft. 1.

GAUCHER (E) : " Culture des bactéries de la lépra." Gaz. méd. de Paris, No. 25.

GAUCHER UND HALLAIRET : " Untersuchen über den Lepra Parasiten." Soc. de Biologie, 11 Dec., 1880. Revue de Med. 1881, p. 71.

HANSEN (AM.) : " *Bacillus Lepræ.*" Virchow's Archiv. Bd. 79, also Quarterly Journal of Microscop. Sci. Jan'y, 1880. II. "Studien über Bacillus Lepræ." Virchow's Archiv, T. 90, p. 542.

KOBNER (H.) : " Uebertragungsversuche von Lepra auf Thiere." Virchow's Archiv. Bd. 88. p. 282.

KOCH (R.) : Mittheilungen aus den Reichs. Gesundheitsamt. p. 10. 1881.

MEISSEN: "Zur Aetiologie der Lepra." Virchow's Archiv. Bd. 84.

MORETTI (O.): "Il primo caso di lebbra nelle Marche confermato dalla presenza del *Bacillus leprœ*." 8°. 10. pp. con. 4 taf. (Bologna) Milano (L. Vallardi) 1883.

MULLER (F.): "Ein Fall von Lepra." Deutsches Archiv. f. klin. Med. 34, p. 205.

NEISSER (Dr. Albert): "Weitere Beiträge zur Aetiologie der Lepra." Virchow's Archiv. Bd. 84. 1881. p. 514–543. Taf. XII., Also in Breslauer Aerzt'l. Zeitschr. 1879. No. 20–21.

THIN: "Reports on leprous infiltration of the epiglottis, and its dependence upon the *Bacillus leprœ*." Brit. Med. Journ. 1884. July 19.

MALARIA.

Bacillus malariæ (Klebs).

Klebs and Tommassi-Crudelli's Method. Infectious material was obtained from the air of malarial districts by the use of a fan ventilator worked by hand. This instrument brought large volumes of air with great force against a glass plate standing at right angles to the air current; this plate was covered with glycerine gelatine, in which the particles contained in the air fixed themselves (the germs from stagnant water did not seem to be the carriers of malaria). From the material obtained from the air, fractional cultures were made in fluid media of various kinds. Some of these pure cultures were then filtered through gypsum and other filters, and inoculation experiments made. Animals inoculated with the unfiltered cultures were, as a consequence, sick with typical malarial fever, while those inoculated with the filtrate showed only a slight rise of temperature, and no periodical rise.

The various forms were studied microscopically in air chambers coated with white of egg, and by staining with anilin-violet, according to general rules.

For the study of earth, they placed upon a sandbath a porcelain vessel having a large surface; this was filled with the damp earth in layer of about 5 cm. thick, and was kept moist by the addition, from time to time, of a little water. Upon this prepared earth was placed a metal box containing the earth to be examined; the bottom of this box was pierced with numerous small openings. The whole apparatus was then kept at a temperature of 30°–35° C. Many of the organisms present in the earth (diatoms, desmids, Hypomycetes, etc.) died, and were eliminated without any appearance of putrefaction. Fractional cultures were then made from the earth contained in the box, and animals inoculated from the pure cultures. Clay cylinders or plaster of Paris cylinders were used as filters, but it was found that Swedish filter-paper would also keep back the *Bacillus malariæ*.

Richard's Method. Richard followed in part the procedure of Laveran, which consisted of direct examination, without any additional fluid, of the blood taken from a patient's finger by pricking. It is necessary to take a very small drop of blood in order to have the blood corpuscles well separated one from another. This method is the only one to be employed for studying the microbe and its movements; but it is insufficient, as it does not disclose the parasitic corpuscles in blood which contains but few. To bring them out under these circumstances he proceeds to destroy the normal red corpuscles, by mixing with the drop of blood a drop of acetic acid. The parasites are not destroyed, but are found with great facility. This method has the advantage that it preserves for some time the elongated parasitic corpuscles, and them only. This microbe has a special habitat, according to Richard, the red blood-corpuscles in which it develops itself, " comme un charançon dans un lentille et d'ou il sort une fois qu'il est arrivé à l'état parfait."

LITERATURE.

BOCHMANN: " Zur Lehre von der Febris intermittens," Vorlaufige Mittheilung, Centralbl. f. d. med. Wissensch. No. 33, 1880.

CECI (A.): " Del germi ed organismi inferiori contenuti nelle terre malariche e commune," 4° 118 pp. (Roma), Milano (L. Vallardi), 1883. II. "L'action de la Quinine (hydrochlorate) en rapport avec le developpment du germs et des organismes inférieurs qui trouvent dans les terres malariques (Malaria)." Trans. Internat. Med. Congress (London). Materia Medica, Rome, 1882.

CUBONI AND MARCHIAFAVA: "Neuen studien über die Natur der Malaria" ("Nuovi studj sulla natura della malaria," Roma, 1881), Archiv. f. experim. Pathol., Bd. XIII., p. 265.

KLEBS (EDWIN) and TOMMASI-CRUDELI: I. "Sulla natura dell' agente specifico che produce le febbri da malaria," Reale Academia dei Lincei, Vol. III., Ser. 3, del 1 gingno, 1879, also in Archiv. f. experim. Pathol. u. Pharm., Bd. XI., p. 126 (chap. II. on methods of investigation); II. "Einige Sätze über die Ursachen der Wechselfieber und die Natur der Malaria," Ibid. Bd. XI., Heft 2, p. 122, also Heft 5 and 6, p. 311; Tab. II.–VI., III., "Studien über die Ursache des Wechselfiebers und der Malaria." Ibid. Bd. XII.

LANZI AND TERRIGI (attribute the cause of malarial fever to a Bacterium which appears inside of a green alga): Centralbl. f. med. Wiss., 1876, No. 40.

LAVERAN (A.): "De la nature parasitaire des accidents de l'impaludisme." Compt. Rend. T. 93 (1881), p. 627–30. Rev. Intern. Sci., IV. (1881), p. 459–61.

MARCHAND : "Kurze Bemerkung zur Aetiologie der Malaria," Virchow's Archiv., Bd. 88, p. 104.

RICHARD : I. "Bestätigung der Laveran'schen Beob. über d. Malaria-parasiten," Compt. Rend. No. 8, 1882 ; II. "Parasiten der Malaria," Ref. in Archiv. der Pharmacie 220, p. 463.

ROZSAHEGGI (A.) : "Von der Ursache des Wechselfiebers," Biol. Centralbl., Bd. II., 1881, p. 97.

SEHLEN : "Studien über Malaria." Fortschritte der Medicin. 1884. No. 18.

STERNBERG (G. M.) : "On the Aetiologie of Malarial Fever." National Board of Health Bulletin, Supplement, No. 14, Washington, 23 July, 1881, 4 pl.

TOMASSI-CRUDELI (CONRAD) : I. "Der *Bacillus Malariæ* im Erdboden von Selinunte und Campobello," Archiv. f. experim. Pathol. u. Pharm., Bd. XII., H. 3, p. 225 ; II. "Die Malaria von Rom," Deutsch von Dr. A. Schaste, M. Rieger, Munchen, 1882 ; III. "Die Malaria von Rom und die alte Drainage der römischen Hügel," Munchen, 1882, 30 pp.

ZIEHL : "Einige Beobachtungen über den *Bacillus Malariæ* (Klebs)." Deutsche med. Wochenschr., No. 48, 1882.

MALIGNANT ŒDEMA.
Bacillus œdema (Koch).

This is the same form which Pasteur names "*Vibrione septique*," or "*Vibrione pyogénique*," the latter being the microbe of puerperal fever.

Heiberg and Laffler associate *Micrococci* with puerperal fever. There is evidently room for a series of careful experiments in connection with this disease.

Malignant œdema is also known as "Pasteur's septicæmia," and is distinct from "Davaine's septicæmia," which is caused by a *Bacterium*.

According to *Flugge*, the *Bacilli* of malignant œdema have, when stained with the aniline dyes, a somewhat characteristic granular appearance.

Pasteur's Method in puerperal fever was to cultivate blood taken from the finger of the patient in chicken broth (culture experiments with the fetid lochia were without result). Rabbits and sheep, into the abdomen of which five drops of the cultures had been introduced, developed within forty-eight hours immense abscesses, which opened spontaneously. Cultures made from the milk of the mother gave like results.

LITERATURE.

BRIEGER and ERLICH: "Ueber das Auftreten des Malignen œdems bei Typhus abdominalis,"

Berliner klin. Wochenschr. No. 43, 1882, p. 661.

FLUGGE: "Fermente u. Mikroparasiten" in Zeimssen's Handbuch der Hygiene, Leipzig, 1883, p. 126.

HEIBERG: (On puerperal fever). Leipzig, 1873.

KOCH (R.): "Bacillus d. Malignen Œdems," Mittheil. a. des. kais. Gesundheitsamt, 1881.

LAFFLER (TH.): "Gehirnerweichung bedingt durch Mikrokokken, Infection bei puerperaler Pyæmia." Breslauer arztl. Zeitschr. 1880, p. 205.

MAYERHAFER: (Vibrio of puerperal fever) Monatsschrift fur Geburtskunde, XXI.

PASTEUR (L.): "De l'extension de la theorie des germs à l'etiologie de quelques maladies commune." Compt. Rend. T. 90, p. 1033 (1883) II. "Ueber d. Vibr. septique," Bull. de l'acad. de med., 1877.

SEPTICÆMIA OF DOMESTIC MOUSE.
Bacillus septicus (Koch).

This form of septicæmia is also known as Koch's septicæmia.

Koch's Method of cultivation was by solid media, either a mixture of aqueous humor and gelatine, or of gelatine, peptone (1 per cent), salt (0.6 per cent), and sodium phosphate in sufficient quantity to render the mass alkaline in reaction. The *Bacilli* grow well on this mixture, and by repeated and rapid division form peculiar branched series.

SEPTICÆMIA OF RABBITS.
Bacterium septicæmiæ (Koch).

Koch produced this disease in rabbits by injecting them with water from the rivulet Pauke, and with putrid mutton infusions. The microbe seems to be the same as that of "Davaine's septicæmia," which was first produced by injecting rabbits with putrid ox blood. The two diseases are distinguished in that Davaine's septicæmia is easily transmissible to guinea-pigs, but not to birds; while mice, pigeons, fowls, and sparrows are very susceptible to the disease discovered by Koch, but guinea-pigs, dogs, and rats resist. Koch says that upon examining more closely the "new disease" described by Pasteur, as being induced by a child dead of hydrophobia, it was found to be "the already long-known septicæmia of rabbits." This should, however, be further investigated.

This *Bacterium* stains in such a manner that between the intensely colored poles a portion, about one half of the entire length, remains unstained. This gives the microbes an apparent dumb-bell shape, and they are easily mistaken for two micrococci.

LITERATURE.

DOWDESWELL (G. F.) : "On the action of heat upon the contagium in the two forms of septicæmia known respectively as 'Davaine's' and 'Pasteur's.'" Proc. Roy. Soc., 1882. No. 221.

GAFFKY : Mittheilung aus dem Reichs-Gesundheitsamt., Bd. 1, 1881, p. 97.

KOCH (R.) : "Bacillus d. Septicæmie bei Mausen." *Ibid.*

SYPHILIS.
Bacillus Klebsii.

Syphilis furnishes another example of a disease having two claimants to the position of *materias morbi*, a *Micrococcus*, and a *Bacillus*.

Birsch-Hirschfeld's Method: Make sections of syphilitic lesions which have been hardened in absolute alcohol. Stain these in a concentrated aqueous solution of fuchsin, obtained by thinning an alcoholic solution of the dye; wash in distilled water; dehydrate with absolute alcohol; clear with oil of cloves, and mount in Canada balsam.

Klebs' Method of inoculation was to cut out pieces of syphilitic lesions, using all precautions, carbolic acid solution spray, etc.; pure cultures were then made from these pieces, and apes were inoculated with the cultures. These animals showed typical syphilitic lesions as a result of such inoculation.

LITERATURE.

AUFRECHT: "Ueber den Befund von Syphilis-Mikrokoken." Ctbl. f. d. med. Wiss, 1881. No. 13, p. 228.

BIRCH-HIRSCHFELD (F. N.): "Bakterien in syphilitischen Neubildungen." Ctbl. f. d. med. Wiss. 1882, No. 33. Abstr. in Deutsch. med. Wochenschr. 1882, p. 505.

HEYDEN (W. H. VAN DER): "Preservation de la syphilis par la vaccin. Traitement des maladies infecticuses, Hypothèse sur le rôle des microbes dans la formation des animaux." Transl. from Dutch by A. E. Roberts, 8° IV., 51 pp., Utrecht (J. L. Beijers), 1883.

KLEBS (E.): "Das contagium der Syphilis. Eine experimentelle Studie." Archiv. f. Exp. Pathol., etc., X., p. 161. II. "Bakterien in frisch extirpirten syphilitischen Primarindurationen." Archiv f. exper. Pathol. Bd. X. Hft. 3 u. 4, p. 161–221.

LUSTGARTEN (S.): "Ueber specifische Bacillen in syphilitischen Krankheitsproducten." (Vorl. Mitt.) Wiener med. Wochenschr, 1884. No. 49.

MARTINEAU ET HARMONIE: "De la bacteridie syphilitique; de l'evolution syphilitique chez le porc." Gaz. hebdomad. 1882, 8 Sept., p. 589. Abstr. in Deutsch. med. Wochenschr., 1882, p. 670.

MORISON: I. "Ueber das Vorkommen von Bakterien bei Syphilis." Wiener Med. Wochenschr. 1883. No. 3.—II. "Ueber das Vorkommen von Bakterien in syphilitischen Secreten." Prager med. Wochenschr. 1883. No. 13.

STRAUS (I.); "Sur la virulence du bubon qui accompagne le chancre mou." Compt. Rend. T. 99, Nov. 24, 1884, p. 935.

TUBERCULOSIS.
Bacillus tuberculosis (Koch).

The property of retaining their color against the action of strong acids, after having been once stained, is characteristic of tubercle *Bacilli*, and of none other, so far as is known, with the single exception of those of leprosy. We may be certain of our diagnosis when we find the *Bacilli* retaining their color after being exposed to the action of acid solutions. It is often the case that the stained tubercle *Bacilli* show uncolored portions of a round or oval form, which are regarded as spores. I have given below, for the sake of completeness, all the so-called "new methods," which are mostly only modifications of Ehrlich's method, which for diagnosis answers every requirement, and is accepted as the best by Koch, Friedländer, etc.[1] Dr.

---

[1] In the employment of any of the methods given for sputum, it is very important that the layer of sputum should not be too thick, — about such a layer as one uses when studying blood. Make about six sputum preparations at once.

Koch originally held that tubercle *Bacilli* would only take staining when simultaneously exposed to the action of an alkali. His original method of procedure was as follows : —

Koch's Method. — The material to be studied is prepared according to general rules for " cover-glass preparations," or sections, and placed for from twenty to twenty-four hours in an alkaline methyl-blue solution [F. 23.]. The time may be shortened, if desirable, to one hour by heating in a drying-oven at 40° C. Wash the stained preparations in water, and place them in a filtered concentrated aqueous solution of vesuvin for one or two minutes for sputum, and fifteen to twenty minutes for sections, then wash again in distilled water until all blue color disappears and a more or less deep brown color is left. Under the microscope all portions of the tissue, especially the nuclei, and any micrococci or products of decomposition which may be present, will be found stained brown, while the tubercle *Bacilli* are a beautiful blue. Dr. Koch says: "All other *Bacilli*, except lepra *Bacilli*, which I have investigated, take a brown color in this method." After staining, dehydrate with absolute alcohol, clear up with oil of cloves, and mount in Canada balsam. The above method is, however, an unsatisfactory one, since the *Bacilli* are hard to find, and the sputum preparations are not good on account of the solubility of the mucin in the staining fluid. It was improved, and a good substitute found for the caustic potash in the so-called "aniline oil" [F. 4.] in —

Ehrlich's Method (now recommended by Dr. Koch), which was based upon the hypothesis that the *Bacilli* were penetrable to staining fluids having a low alkaline reaction, but that they were surrounded by an envelope impenetrable to acids. Erlich sustained this opinion by subjecting his preparations to a solution of nitric acid, by which the tissues were entirely bleached, while the *Bacilli* retained their color.

For sputum he pressed a particle between two cover-glasses by means of a preparation needle, and, *sliding* them apart, passed them, prepared surface up, through the gas flame. He then allowed them to float, prepared side down, in a watch-glass filled with gentian-violet, methyl-violet, or fuchsin-anilin-oil solution [F. 17.] for fifteen minutes to half an hour. By heating the whole over the flame until it steams, they need only be left for one minute. Wash in a solution of nitric acid (33⅓ per cent), under the influence of which the color soon fades out in the matrix, the *Bacilli* alone retaining the color — violet or red. Wash in distilled water, dry, pass through absolute alcohol, and mount in Canada balsam. A better and clearer picture is obtained by making a double staining, *i. e.*, re-staining the decolorized matrix. This is best accomplished by means of some complementary color, *e. g.*, aqueous solutions of Bismarck-brown, vesuvin, or malachite-green. Ehrlich's method excelled that proposed by Koch in its greater rapidity and in the large number of *Bacilli* colored. In the discussion following its announcement Dr. Koch said he preferred it to his own, and now uses it altogether. Its disadvantages are, however, that many *Bacilli* are decolorized.

For sections. — Thin sections must remain in the staining fluid for twenty-four hours, and for two or three minutes in the acid mixture. Wash well in repeatedly-renewed water (hereupon depends the tenacity of the *Bacilli* staining), pass through absolute alcohol, oil of bergamot or cloves, and mount in Canada balsam. Some months after Ehrlich published the above modification of Koch's first method, Ziehl opposed the statement that the external envelope of tubercle *Bacilli* is only penetrated by coloring matter when under the influence of alkalies, and substituted, for aniline oil and caustic potash, carbolic acid, according to the following plan: —

Ziehl's Method. — I. Prepare as in Ehrlich's method, but omit the decolorizing with the nitric acid solution, which drives out the coloring matter from all the Schizomycetes with the exception of the tubercle *Bacillus*, thus allowing the working of the methyl-blue solution. II. Prepare exactly as in Ehrlich's method, but in place of the aniline oil use resorcin, pyrogallic acid, or carbolic acid. In this manner he demonstrated that the tubercle *Bacilli* did not need an alkaline staining fluid.

Still later Dr. Prior showed that — (1) Ehrlich's staining fluid is not alkaline, but neutral; (2) that the staining of tubercle *Bacilli* is successful if a neutral, alkaline, or weakly acid fluid be used; (3) that it is successful if a non-alkaline fluid, *e. g.*, turpentine oil, takes the place of the aniline oil. The next modification proposed was the —

Balmer and Fränzel'z Method. — Prepare as for other methods, and allow the cover-glasses to float for twenty-four hours, prepared side down, upon a quantity of aniline-gentian-violet solution [F. 19*a*]. Then wash in acidulated water [F. 7.] for one half to one minute, or until the color disappears, wash in distilled water, and color the ground with a concentrated aqueous solution of Bismarckbrown [F. 2.]. This method has the advantage that few *Bacilli* are overlooked, but it takes too long, and is inconvenient, as it requires the staining fluid to be made new each time.

In view of this inconvenience, and to afford a method which physicians could use to advantage for diagnosis, Rindfleish taught his pupils the following method : —

Rindfleish's Method. — Pour into a test-tube sufficient aniline oil to fill the fundus, then add water until it is a third full, shake it well, and filter through a small filter, that can be held in the hand, into another test-tube. To the clear filtrate add eight drops of a concentrated alcoholic solution of fuchsin [F. 16.]. Now place before you upon a piece of white paper — (1) a watch-glass half full of alcohol, to which add two drops of dilute nitric acid; (2) another watch-glass half full of the above prepared fuchsin solution; (3) a lighted spirit lamp. Now with the pincers take up the cover-glass holding the dried sputum, and pass it three times, with the sputum side up, through the flame ("about as fast as you would cut bread"), thus rendering the albumen homogeneous. Now lay the cover-glass, prepared side down, upon the staining fluid, and with the pincers pick up this watch-glass and hold it over the flame until it begins to steam, then remove the cover-glass, and after washing it in a stream of distilled water, place it in the acidulated alcohol, allow it to remain here until the violet clouds fade out. In from ten to fifteen seconds it will appear to be stained only at a few points; now wash it in distilled water, dry it, pass it through absolute alcohol, oil of cloves, and mount in Canada balsam. For examination use your highest powers and remove diaphragms.

Orth's Method was to use water acidulated with hydrochloric acid [F. 7] in place of the usual nitric acid water, and to stain with picro-carmine in place of vesuvin. This gives the *Bacilli* blue, and the nuclei red. Stain first in gentian-violet, wash in distilled water, transfer to picro-carmine, then to hydrochloric acid alcohol; wash in pure alcohol, and mount in damar. Never use Canada-balsam dissolved in chloroform.

The next method is not a good one for diagnostic purposes, because the peculiar relation of tubercle *Bacilli* to coloring-matter is not based upon taking on of color, but only upon the power which they possess of retaining, to a certain degree, this color, when it is once taken on, against the effect of mineral acids. If one carries on the decolorizing process under the microscope, he will note that upon the addition of nitric acid all the color disappears from the *Bacilli*, but that upon the addition of water it returns.

Lichtheim's Method. Prepare according to general rules. The staining fluid used is a concentrated aqueous solution of fuchsin or gentian-violet, made by adding some of the alcoholic solution to some distilled water. Allow the cover-glasses to float in this for thirty-six hours. The time may be shortened by warming. The *Bacilli* do not stain so strongly as by other methods; wash in water, absolute alcohol, oil of cloves, and Canada balsam. The next modification is —

Petri's Method. I. Petri, like Ziehl, considered the use of aniline oil as superfluous. Prepare clear filtered solutions of fuchsin and malachite-green in alcohol, thin 5 ccm. of each, with 100 ccm. of distilled water. The cover-glasses are prepared according to general rules. Cover the fuchsin solution with a glass plate, and heat until the cover is dimmed with condensed vapor, then allow the prepared cover-glasses to float, prepared side down, upon the hot fluid, and cover it again, allowing it to stand for from one to fifteen minutes. Then take out the deeply stained preparations, and place them in a glass dish having a lip; pour over them some glacial acetic acid; in a few minutes they will be sufficiently decolorized. If the acetic acid is made very red, wash them a second time in some fresh acid. Then wash them three or four times with water, by decanting, in the same dish, and finally pour the malachite-green solution over them. After from five to ten minutes they will be sufficiently stained. Now wash in water, dry upon filter-paper placed on a warm surface. Mount in glycerine or Canada balsam, and examine with a hydrate of chloral immersion objective. This is better than an oil immersion, because the preparation is easily washed by pouring water over it. The *Bacilli* will be stained red, the other parts of the sputum blue-green, and they keep color better than when mineral acids are used.

II. Instead of fuchsin and malachite-green, aniline-gentian-violet [F. 19] and aniline yellow [F. 6] may be used. Allow the sections to remain in a fresh solution of the aniline-gentian-violet for half an hour, then for eighteen hours in 20 ccm. of absolute alcohol, rinsing two or three times; wash for one minute in distilled water, and place in an aqueous solution of aniline yellow for three minutes. Wash in absolute alcohol, pass through oil of cloves, and mount in Canada balsam. The advantages of this method are the large number of *Bacilli* stained, and the permanence of the staining. Dr. Plaut, of Leipzig, had preparations made by this method remain good for over a year, and recommends it for dealers in microscopical preparations.

Gibbs' Method. Prepare according to general rules. Allow the preparations to remain for from fifteen to twenty minutes in a solution of magenta [F. 26]. Then wash in an aqueous nitric acid mixture (1 part to 3), and afterwards in alcohol. Give a double color with chrysoidine [F. 11]. Wash in alcohol, pass through oil of cloves, and mount in Canada balsam.

Gibbs' New Method. Allow the prepared cover-glasses to float for five minutes in a solution of Rosalin-chlorhydrate methyl-blue [F. 36. 23], which has been warmed to steaming. Wash in methylic alcohol until no more coloring matter is given off. Dry; pass through absolute alcohol, oil of cloves, and mount in Canada balsam.

Sections must remain several hours in the staining fluid, otherwise they are handled the same as cover-glass preparations. The advantages of this method are, that on account of no acids having been used there is no shrinking of the tissue, and the preparations retain their color very well. Accidental Bacteria, in the neighborhood of the *Bacilli*, will often be found stained red.[1] The above is a method which is strongly recommended, as is

Baumgarten's Method (for sputum). Prepare according to general directions for cover-glass preparations, and moisten the dried preparations in a watch-glass of distilled water, to which has been added one or two drops of a solution of caustic potash. By the use of a power of four or five hundred diameters the *Bacilli* are already recognizable. They do not appear any larger than those prepared by the aniline coloring methods. To avoid confusing them with other Schizomycetes, drop upon the dried cover-glass a drop of aqueous aniline-violet solution. The tubercle *Bacilli* appear absolutely colorless. This process does not give good results for diagnosis, but is a convenient method of rendering the *Bacilli* visible

---

[1] Drs. Baumgarten and Fränkel claim that frequently the accidental Bacteria take the blue color, while the *Bacilli* are stained red, thus rendering the method useless for diagnosis. — Vid. Zeitschr. f. wiss. Mikroskopie, Bd. 1, Heft. 3, p. 457.

without coloring, and is recommended especially for demonstration, as it shows that the tubercle *Bacilli* behave differently from most other forms of *Bacilli* in being decolorized, as well as the nuclei of the tissue. This also happens if they are prepared according to Gram's general method.

Baumgarten's New Method. In a small watch-glass full of distilled water drop four or five drops of concentrated alcoholic solution of methyl-violet. Allow the preparations to become deeply stained in this fluid; wash in water for from five to ten minutes, and pass through absolute alcohol; then lay them in a half-saturated solution of carbonate of potassium, whereby the tissue is completely decolorized. Wash in absolute alcohol, pass through oil of cloves, and mount in a mixture of equal parts of oil of cloves and Canada balsam (free from chloroform).

(For sections.) Stain as for cover-glass preparations, then place for five minutes in alcohol, and afterwards in an acetic acid solution of Bismarck-brown; finish as for cover-glass preparations. accidental bacteria appear, according to this process, brown; if numerous, covering up the often sparsely present tubercle *Bacilli*. The use of the carbonate of potash solution is also to be recommended here, before using the Bismarck-brown. In place of the latter, borax carmine or borax-picro-carmine may be used.

Baumgarten's Culture Method.—Take from a living animal a small fragment of a Tubercle nodule (or from a perfectly fresh human nodule, rich in *Bacilli*), and introduce it with all the proper antiseptic precautions into the anterior chamber of the eye of a living rabbit. Here the piece increases in size, owing to the development of the *Bacilli.* Allow it to remain for six or eight days, and then remove it and take a small particle from it. Place this second particle in the anterior chamber of the eye of another living rabbit, allow this to remain six or eight days, and place a third particle in the eye of a third rabbit. Proceed in this way until an absolutely pure culture is obtained. This method has the advantage over the methods of pure culture outside the living body, with artificial culture apparatus, that it requires, with the exception of the antiseptic precautions of the operation, no further trouble with sterilization or regulation of temperature.

Weigert's Method. Prepare according to general rules, and stain in gentian-violet solution [F. 18.], and handle further according to Ehrlich's method. The advantages of this method are, that it renders visible the Bacteria of a section when present in very slight numbers, and is to be recommended for use in tuberculosis of men and cattle.

Fränkel's Method, for the rapid double staining

of sputum. To 5 ccm. of aniline-water [F. 5.], heated to 100° C., add, drop by drop, a saturated solution of methyl-violet or fuchsin, until a strong opalescence is produced. Allow the cover-glasses to float upon the heated fluid for two minutes, and then decolorize, and re-stain simultaneously in an (herein lies the peculiarity of the method) acid alcoholic solution of a contrasting, complementary dye. (1.) For blue, use alcohol 50 parts, water 30, hydrochloric acid 20, and as much methyl blue as will dissolve after repeated shaking; filter. (2.) For brown, alcohol 70 parts, hydrochloric acid 30, as much Vesuvin as it will dissolve; filter. (3.) For green, alcohol 50 parts, water 20, acetic acid 30, and as much malachite or methyl-green as it will dissolve; filter. Lay the stained cover-glass in one of these prepared solutions (which may be kept on hand) for one or two minutes, wash in water or dilute acid (1 per cent acetic), then in 50 per cent alcohol, dry, first between blotting-paper, then over the flame. In four minutes a good double-stained preparation is made, which is permanent.

Pfuhl Petri's Method. Prepare according to general directions, and stain in a solution of fuchsin [F. 16.] by floating the cover glasses in it for one or two minutes, warm to steaming in the meantime. Decolorize in acetic acid, wash in water. Give a second staining by means of a solution of malachite green [F. 24] thirty seconds to one minute, wash in water, pass through absolute alcohol, oil of cloves, and mount in Canada balsam.

Senkewitsch's Method. Prepare according to general directions, and place in a concentrated solution of fuchsin. After the preparations are sufficiently stained, wash for one or two minutes in 10 cc. of alcohol, to which one drop of nitric acid has been added. Wash in water, dry, pass through absolute alcohol, oil of cloves, and mount in Canada balsam.

Käätzers (combined) Method. Spread out a quantity of sputum upon a black plate, and pick out the opaque white or gray-white patches by means of two previously sterilized needles. Prepare cover glasses according to general directions, and stain in a filtered solution of aniline-gentian-violet [F. 18] by allowing the cover-glasses to float for twenty-four hours or more in the same, or hasten matters by heating up to 80° C. for several minutes, remove and soak up the excess of staining fluid on the cover glasses by means of a bit of blotting paper, then place in a watch glass of acidulated alcohol [F. 7] for from one half to one minute, then remove to a watch-glass containing 90 per cent alcohol, until the color has faded out, then wash in distilled water. Now drop upon the preparation four or five drops of a concentrated aqueous solution of vesuvin. After two minutes, wash in distilled water, dry, pass through absolute alcohol, oil of cloves, and mount in Canada balsam or damar varnish.

Long's Method for Sputum. The entire quantity of sputum to be investigated is placed in alkaline water. For this purpose a watch-glass is used, into which is poured five or six grammes of distilled water, and to this are added three or four drops of a 33 per cent caustic-potash solution by means of a burette; after this is well mixed the sputum is placed in it. In about half an hour the latter is much dissolved, the air-bubbles much diminished, and grayish-green stripes may be seen in the compact masses, which, if *Bacilli* are present, are sure to contain them. By this method small clear-white scales are seen which resemble the culture colonies as described by Dr. Koch, which they really are. Stain according to Balmer-Fränzel's Method, and use for demonstration an Abbe's condenser, and a one-twelfth homogeneous immersion objective.

Peter's Method for Sections, to stain all the bacilli present. Thin sections, which have been hardened in alcohol, are placed for one moment in distilled water, then for thirty minutes in a two per cent aqueous gentian-violet solution (filtered and made with fresh aniline-water). Then place for eighteen hours in at least twenty grammes of absolute alcohol in a closed vessel, renewing the alcohol once or twice. Wash for one minute in distilled water, then place for three minutes in a two per cent aqueous solution (filtered) of aniline-yellow, then from five to thirty minutes in absolute alcohol. By the shorter use of absolute alcohol the *Bacilli* are easier found, and the method of grouping more apparent; by the longer use the structure of the tissue surrounding the *Bacilli* is made more evident. Pass through oil of cloves and mount in Canada balsam.

Veraguth's Method. This method is a modification of Ehrlich's, proposed in order that the large number of pathological preparations in collections might not be lost. By its use tubercle *Bacilli* can be shown in old chromic acid preparations as readily as in freshly hardened alcoholic preparations. Lay small pieces of the preparation to be examined for two or three days in flowing water, and then place in alcohol. When ready to cut, place them for twenty-four hours in water, and then cut with a freezing microtome. For pieces of lung, use with this instrument a solution of gum arabic in glycerine. Before staining, place the sections for twenty-four hours in absolute alcohol, and then stain in aniline-water-fuchsin for forty-eight hours. Decolorize with nitric acid water, and wash well in distilled water. Give double-staining with an aqueous solution of methyl-blue for from five to ten minutes, wash in water, pass through absolute alcohol, oil of cloves, and mount in Canada balsam.

Coze and Simon's Method for Sputum. — The process is the same for the examination of all liquids, only when the liquids are insufficiently albuminous to adhere to the glass, a few drops of fresh white of egg is added, which gives, by its coagulation, more fixity to the preparation. Prepare cover-glasses according to general directions, and place in a solution of fuchsin or gentian-violet. (F. 19, b.) Heat to 40° C. for half an hour, wash in water, pass through acid solution (F. 7.), wash again in plenty of water, and place for two or three minutes in a solution of chrysoidin (F.), wash dry and mount.

For Sections. — Harden the pieces of tissue in absolute alcohol for forty-eight hours or more, cut extremely thin sections, and place them for twenty-four hours in a solution of fuchsin (F.), or for one hour if temperature is raised to 40° C. Wash 4–5 minutes in acid solution (F.), then in water until the last trace of acid is removed. Then place for 15–20 minutes in the aniline water (F. 7.) and afterwards stain in chrysoidin or hæmatoxylin (F. 11.22.). Wash, dehydrate in absolute alcohol, pass through turpentine, and mount.

Déjérine's Method for studying the stony concrements which are found, about the size of a pea, in the apices of the lungs of old people, was to pound them up in a mortar with a little distilled water. In the limy broth he found fragments of organic substance which he stained after Ehrlich's method, and in which he found tubercle *Bacilli*. These concretions have a hard nucleus and a peripheral zone of the consistence of poorly hardened gypsum.

Giboux's Method, to test the inoculability of tuberculosis through respiration, was to use two boxes, in which were placed rabbits, and through which he passed daily 20,000–25,000 cm. of air that had been exhaled by phthisical patients. One box was protected by having the air filtered through cotton. After three and a half months the rabbits in the unprotected box died, and showed in their lungs, livers, and spleens, tubercle *Bacilli*. The other rabbits remained healthy.

Ermengen's Method. — Add the chosen aniline dye to a mixture, of aniline-oil 4 grammes, alcohol 20 grammes, at 40° C., to this add an equal quantity of distilled water. Prepare fresh, and filter before using. Sulphate of rosanilin and methyl-violet, quality BBBBB, preferred. Decolorize the preparations in dilute nitric acid, and wash thoroughly in distilled water. For a second stain, use aniline-blue, aqueous solution of vesuvin, or, still better, 'Grenacher's carmine' when the staining of the *Bacilli* has been effected by means of methyl-blue or -green. Avoid exposing the preparations to a bright light. Mount in Canada balsam, or damar dissolved in benzine, which preserves well the colors produced by the aniline salts. Glycerine-jelly may also be used successfully.

Brun's Method.— Spread out by pressure between two cover-glasses a particle of sputum, and allow it to remain exposed to the air, but protected from dust, for a few minutes. The thin, mucilaginous layer thus obtained contains albumen, mucin, pus granules, fat and blood cells, particles of carbon, and other debris of respired air. Drop on the prepared cover-glass a few drops of a concentrated solution of fuchsin or methyl-blue, (made with equal parts of water and alcohol containing 1-30th aniline oil to render it alkaline). Drop the cover-glass into a test tube and wash in water by decanting, then wash in an acidulated solution, (F. 7.), until the organic matrix is very feebly colored; wash repeatedly by decanting in the same test tube; then cover the preparation for a few minutes with a concentrated solution of aniline oil; this neutralizes all remaining acid. Mount in fluid (F. 43). Keep the preparations in the dark. The advantages claimed for this method are that:—(1) It avoids heating up to 100° or 120° C., which coagulates the albumen and renders it opaque, besides diminishing the size of the microbes by contraction. (2) The organic matter is rendered transparent by means of acetic acid. (3) The action of the nitric acid remaining in the organic matrix is neutralized and prevented from decolorizing the Bacteria and rendering them invisible after a time, as occurs very often in other methods. (4) Instead of Canada balsam, which has a very high refractive index (1.53), a neutral liquid having the same refractive index (1.37) as the albuminoid substances, is used.

Burrill's Method. — Not finding any of the alcoholic solutions of aniline dyes satisfactory, on account of the formation of a precipitate on the cover-glasses during staining, Prof. B. substitutes a solution of fuchsin in glycerine (F. 16). Place the cover-glass in this solution for from ten minutes to several days, at ordinary temperatures. The best results are obtained by heating up to 80° or 100° F. for 25–30 minutes. If it is desired to make haste, which gives less good results, boil a little water in a test tube, add gradually to this double its quantity of the above staining fluid, which is allowed to flow down the side of the tube held obliquely, shake carefully, turn the fluid into a watch-glass, and plunge into it the prepared cover-glass, allow it to remain for 1–2 minutes, decolorize in a solution (1–4) of nitric or hydrochloric acid, wash in water, dry, and mount.

Hartzell's Method. — Prepare cover-glasses according to general directions. Place upon the dried sputum one or two drops of the fuchsin solution recommended by Gradle (F. 16.), and allow it to remain 3-5 minutes, decolorize completely in oxalic acid solution, wash thoroughly in water, dry, and mount in glycerine or Canada balsam. With a power of 500 or 600 the *Bacilli* will appear as brilliant red rods; no staining of the background is necessary. One chief advantage claimed over other methods is, that in the latter the decolorizing agent employed is dilute nitric acid, but this, besides being disagreeable to handle because of its staining and corrosive qualities, is apt to remove the color from the *Bacilli*, unless great care is exercised. Oxalic acid, however, seems to leave the dye untouched.

Quinlan's Method. — Take a particle of the expectoration of an advanced case (the first expectoration of the morning is preferable), mix it with a few drops of a solution of potash by means of a glass rod, until it is homogeneous; transfer a drop of this mixture to a cover-glass, and dry by passing rapidly through the flame of an alcohol lamp. Place upon the dried sputum a few drops of magenta (F. 26.), allow this to remain for twenty minutes; then place it in a mixture of nitric acid 1 part, distilled water 2 parts, until it is bleached, which will be in about five minutes; then wash in distilled water. Drop upon the preparation a little solution of methyl-blue, allow it to remain five minutes, wash, and dehydrate in absolute alcohol. Place a drop of Canada balsam and benzine upon a slide, and lay the prepared cover-glass upon it. *Bacilli* stained scarlet upon a blue ground.

Negri's Method, for staining the spores in the *Bacilli* of tuberculosis.

(1) Powdered carmine . . . . gr. 0.5
   Strong aqua ammonia . . . . cc. 1
   Distilled water . . . . " 30

Place the carmine in a porcelain capsule, mix it up with a little water, add the ammonia and the remainder of the water. Allow the liquid to remain in the air protected from dust, until every trace of ammoniacal odor disappears, this is essential. Keep it at a temperature of about 15° C. for four or five days. Decant the clear liquid and throw away the precipitate.

(2) Alcohol (ordinary) . . . . cc. 100
   Hydrochloric acid (pure) . . . . drops 20
(3) Concentrated solution of picric acid in distilled water, allow an excess of crystals to remain at the bottom.
(4) Liquid (2)
   " (3) of each 15 cc.

To this mixture add drop by drop, stirring all the time, the solution of carmine (1). (Do not add the acidulated alcohol to the carmine, as it gives a precipitate.) Ordinarily no precipitate is formed; if it does form, decant. Add a small crystal of thymol, and preserve in a stoppered bottle. This solution Negri calls (5).

(6) Methyl-violet . . . . O. gr. 7
    Absolute alcohol . . . . cc. 10
    Aniline oil . . . . " 4

After complete solution of the coloring matter add distilled water, 15 cc.

The methyl-violet obtained by this method gives, according to Negri, better results than gentian-violet. It is necessary to have a good methyl-violet, the greater part of the violets ordinarily sold not answering the purposes of the microscopist.

Spread out a very thin layer of sputum upon a cover-glass; do not heat at all, or only very slightly, as too much heat renders the *Bacilli* difficult to stain, over-heating is probably the cause of much failure. Place the cover-glass, prepared-side up, in a watch-glass, and drop upon it the solution of the aniline dye (6), and allow it to remain covered for ½–1 hour or more at a temperature of 15° C. This method is better than to float the cover-glass upon the staining-fluid, because the aniline oil is never perfectly dissolved, and forms upon the top a layer which prevents the staining being perfect and intense. Wash the cover-glass in plenty of ordinary water until all excess of color has disappeared; then plunge it into the acid alcohol (2), until cleared, wash again *and while still wet*, pour on it a few drops of the carmine solution (5), and allow it to stand five minutes. Wash off the excess of carmine, and then wash again in the acid alcohol (2), until most of the color has disappeared; then plunge into distilled water, which is to be renewed twice in the course of 8–10 minutes. Dry, and mount in pure balsam, by heating a drop upon a slide.

Upon examination one sees numerous spores colored azure-blue, enclosed in the transparent envelopes — which are, however, visible if observed with care — of the *Bacilli*, upon a rose-ground. If it is desired to stain the *Bacilli* entire instead of only the spores, wash the preparation after the action of the carmine in distilled water, without passing through the acid alcohol.

Reinstadler's Culture Method for tubercle *Bacilli*. Place a piece of tubercular lung in a sterilized mortar with some glass sand which has been previously heated red hot, and rub the mass thoroughly, or comminute the lung by cutting with scissors. To the material thus obtained add a small quantity of Bergmann's fluid [F. 1.] which has shortly before been thoroughly cooked and cooled. Mix thoroughly the broth thus formed, and filter; seal up for further use. Several test tubes are then cleaned by boiling them, first in nitric and sulphuric acids, and then in alcohol. After pouring out the latter, heat them in the spirit flame, and stop the perfectly dry tubes with a wad of carbolized cotton. Now fill them by means of a sterilized pipette with about 30 cc. of the Bergmann's fluid, sterilize the whole by heat, and, when cool, transplant according to general rules for cultures.

Celli and Guarnieri's Method for ascertaining whether the microbes of tuberculosis were thrown into the air by the breath of consumptives. I. *For the air used by isolated patients.* They (C. & G.) lit a gas jet in the lower part of a large tin cylinder, open at the top and narrowing down below into a cone, into whose apex a small bent tube was soldered, whose free end terminated in a wide cone of tin, into which was introduced a cone of copper open at its apex, and the surface of which, after having been heated red hot, was coated with Koch's gelatine. The warmth of the gas flame caused a current of air to pass over the gelatine surface. By means of this apparatus, experiments were made for twelve nights at various heights in the room, the usual ventilation of which was closed. After each, the gelatine was kept at an even temperature of 30°–40° C. and examined microscopically by Ehrlich's method; also inoculations made. II. *For direct expiration of consumptives.* A large number of patients were required to breathe repeatedly, throughout twenty-four hours, (1) partly upon a small dish of wood, with a concave bottom, coated with Koch's gelatine, and covered during the pauses of the investigation with a watch-glass; (2) partly into a glass tube coated inside with gelatine, and which was only open during the investigation; (3) partly in flasks of distilled water which had been thoroughly sterilized, and through the corks of which passed two small bent glass tubes, the short one being stopped at its free end with cotton, the patient breathed into the longer tube, one end of which dipped into the fluid; the other having a piece of rubber tubing attached, which could be closed with a clamp; (4) partly in a Liebig's ventilator embedded in ice, the entrance and exit tubes being arranged as in (3). The gelatine from these experiments was investigated by Ehrlich's method and by inoculations. The water condensed from the breath of the patients by apparatus (4) cultivated in sterilized media. III. *Sputum* was placed in a retort on a water bath at 34–40° C., and the vapor thus generated was collected in a bulb packed in ice, the water thus obtained was examined by Ehrlich's staining method and also placed in culture media. Air was also collected from sputum by means of an aspirating apparatus. The result in every case was to show that tuberculosis sputum, so long as it is moist, does not give out its specific *Bacilli*.

LITERATURE.

ARLOING (S.) : Nouvelles experiences comparative sur l'inoculabilité de la scrofure et de la tuberculose de l'homme au lapin et au cobaye. Compt. Rend. de l'Acad. de sc. de Paris, T. 99, Oct. 20, 1884, p. 661.

AUFRECHT : "Die Aetiologie der Tuberculose," Centralblt. f. med. Wissensch. 17, 1882, also Deutsche med. Wochenschr., 1882, p. 283.

BABES : "Comparaison entre les bacilles de la tuberculose et ceux de la lepra (Eléphantiasis des Grecs)." Compt. Rend. T. 96, p. 1244 and 1323 (1883).

BALMER UND FRAENTZEL : "Ueber das Verhalten der Tuberkel-bacillen im Auswurf während des Verlaufs der Lungenschwindsucht," Berl. klin. Wochenschr., 1882, No. 45, p. 679.

BAUMGARTEN (P.) : I. "Tuberkelbakterien," Centralbl. f. med. Wissensch., 1882, No. 15, 19, Deutsche med. Wochenschr., 1882, p. 238 ; II. "Beiträge zur Darstellungsmethode der Tuberkelbacillen." Zeitschr. f. wissensch. Mikroscopie, Bd. I., Hft. 1, p. 51 ; III. "Ueber ein bequems Verfahren Tuberkel-bacillen in Sputis nachzuweisen," Ctblt. f. d. med. Wiss., 1882, No. 25. IV. "Ueber das Verhaltniss von Perlsucht und Tuberculose." Berl. klin. Wochenschr. 1880. V. "Ueber ein neues Reinculturferfahren der Tuberkelbacillen." Ctbl. f. d. med. wiss. 1884, No. 22.

BOLLINGER (O.): "Zur Aetiologie der Tuberculose," München (Rieger'sch Univ. Buchhdl.), 1883. II. "Ueber Tuberkelbacillen in Eiter einer tuberculösen Kuh und über Virulenz des Secretes einer derartig erkrankten Milchdrüsen." Bayerisches arztl. Int. Bl. 1883. No. 16.

BOULEY (H.): "La nature vivante de la contagion, contagiosite de la tuberculose," Paris, 1884.

BRUN (J.): "Note sur les meilleurs proceédes pour reconnaitre et faire des préparations microscopiques des bactéries de la tuberculose." Journal de Micrographie, 1882, Vol. VI., p. 500.

BURRILL (T. J.): "To stain Bacillus tuberculosis;" The Microscope, Vol. IV., 1884, p. 6; Journal de Microgr., t. VIII., 1884, No. 4, p. 240.

BURNET (D.): "Sur la tuberculose experimentale." Compt. Rend. T. 93 (1881), p. 447, 448.

CELLI (A.) ET GUARNIERI (G.): I. "Sopra talune forme crystalline che potrebbero simulare il bacillo del tuberculo" (On such crystalline forms as may be mistaken for tuberculosis bacilli), Acad. dei Lincei, 17, gingno, 1883. — The Microscope, Vol. IV., 1883, No. 6, p. 135. — Zeitschr. f. wiss. Mikros., Bd. 1, 1884, p. 590; II. "Intorno alla profilossi del Tuberculosi, studi d'igiene sperimentale" (on the prophylaxis of tuberculosis, studies in experimental hygiene), Arch. per. lc. se. med., Vol. VII., fasc. 3, 1884, p. 233, 3 Pl.

CHYNE (WATSON) : (On Tuberculosis bacilli). Practitioner, April, 1883.

CLARK (J. W.) : "Preliminary note on Bacillus tuberculosis, Koch." Nature, Vol. XXVII., p. 492.

COLIN (G.) : " Sur la transmission de la tuberculose aux grandes ruminants," Compt. Rend. T. 99, p. 1057 (Dec. 15, 1884).

COPPOCK : " On Bacillus tuberculosis," Microsc. News, Vol. III., 1883, No. 28, p. 121.

COZE ET SIMON : " Recherches de pathologie et de therapeutique experimentales sur la tuberculose," Journ. de Micrographie, t. VIII., 1884, No. 4, p. 235.

DÉJÉRINE (J.) : "Recherche des bacilles dans la tuberculose calcifiée et caseo-calcifiée." Revue de Med., 1884, No. 12.

DEMME (R.) : "Zur diagnostischen Bedeutung der Tuberkelbacillen für des Kindesalter." Berliner klin. Wochenschr. 1883, No. 13.

DETTWEILER AND MEISSEN : "Der Tuberkelbacillus und die chronische Lungenschwindsucht, Berliner klin. Wochenschr., 1883, No. 4, p. 97, also No. 8, p. 117.

DOUTRELEPONT : "Tuberkelbacillen im Lupus." Monats. f. prakt. Dermat. II. No. 6.

ERLICH : " Eine neue Methode der Färbung von

Tuberkelbacillen," Gesellschaft der Charité-Aerzte in Berlin. Sitzung von 27 April, 1882, also Berl. klin. Wochenschr., Jan., 1883, p. 13, also Deutsch med. Wochenschr., 1882, p. 267.

ERMENGEN: "Le microbe de la Tuberculose," Revue Mycologique IV., No. 16. II. "Préparation des Bactéries de la tuberculose. Perfectionnements apportés a la méthode de double coloration." Journal de Micrographie, 1882, p. 466.

ERNST: Articles in two August numbers of Boston Med. and Surg. Journ., 1883, gives extensive résumé of the literature pertaining to Koch's discovery of *Bacilli* of Tuberculosis.

ESCHLE: "Tuberkelbacillen in dem Ausflusse bei mittelohr-Eiterung von Phthisikern." Deutsch. med. Wochenschr. 1883. No. 30.

FORMAD (H. F.): "The Bacillus of Tubercule," New York Med. Journ., 1884, Feb. 16, also Amer. M. Microsc. Journ., Vol. V., No. 4, 1884.

FRAENKEL (B.): I. "Ueber die Farbung des Koch'schen Bacillus und seine semiotische Bedeutung für die Krankheiten des Respirationsorgane," Berl. klin. Wochenschr., 1884, No. 13; II. "Zur diagnose des tuberculosen Kehlkopfgeschwurs" *Ibid*. 1883, No. 4, p. 58.

FRANTZEL (O.): "Wie weit können wir das Nachweis von Tuberclebacillen bis jetzt praktische verwerthen?" Dtsch. militairarztl. Zeitschr., 1883, Aug.

GAFFKY: "Ein Beitrag zum Verhalten der Tuberkelbacillen im sputum," Mitt. aus d. kais. Gesundheitsamt, Bd. 2, 1884. p. 126.

GIACOMI (DE) (on staining): Fortschritte der Med., 1883.

GIBBES (H.): I. "On a rapid method of demonstrating the tubercle bacillus without the use of nitric acid," The Lancet, Vol. I., 1883, p. 771, Journ. Roy. Mic. Soc. Ser. II., Vol. III., 1883, pt. 5, p. 764, Microsc. News, Vol. III., 1883, No. 33, p. 248. II. "A new method for the detection of the tubercle bacillus," British Med. Journ., 14, Oct., 1882.

GIBOUX: "Inoculabilité de la tuberculose par la respiration des phthisiques." Compt. Rend. T. 94 (1882), p. 1391.

GROVE (W. B.): A synopsis of Bacteria and yeast fungi, and allied species, Schizomycetes and Saccharomycetes. Appendix B. On the staining of *Bacillus Tuberculosis*. Ehrlich's, Gibbs's, and Prideaux's Methods. 87 figs. 8°. London, 1884.

HARTZELL (M. B.): "A ready method for the detection of the Bacillus tuberculosis," Med. Times, Jan'y 26, 1884, The Microscope, Vol. IV., 1884, No. 5, p. 115, Amer. M. Microsc. Journ., Vol. IV., 1884, No. 4, p. 76, Journ. Roy. Mic. Soc., Ser. II., Vol. IV., 1884, pt. 4, p. 653.

HERON (G. A.): "Ehrlich's method for the

detection of Tubercle *Bacilli* in sputum." Brit. Med. Journ., No. 1137 (1882), p. 735.

HILLER: "Ueber initiale Hämoptœ und ihre Beziehung zur Tuberculose." Dtsch. med. Wochenschr., No. 47, 1882.

IRSAI (A.): "Zur Diagnostik der Tuberculose des Harn-apparates auf Grund des Befundes von Koch'schen Tuberkelbacillen im Harn." Wiener med. Presse, 1884, Nos. 36, 37.

KAATZER (P.): "Die Technik der Sputum Untersuchung auf Tuberkel-Bacillen," 2d Aufl., Wiesbaden, 1884.

KAROP (G. C.): "On a specimen of Bacillus tuberculosis prepared by Dr. Gibbs method," Journ. Quek. Microsc. Club, Vol. I., 1883, No. 1.

KLEBS (E.): "Ueber Tuberculose," Prager med. Wochenschr., 1877, Nos. 42 and 43. (The first to propose the theory that the tuberculosis virus must contain Bacteria, cultivated these in white of egg.)

KOCH (R.): I. "Die Aetiologie der Tuberculose," Verhandl. des Congresses für innere Medicin, 1882; II. "Die Aetiologie der Tuberculose," Berl. klin. Wochenschr., 1882, p. 221, also Deutsch med. Wochenschr., 1882, Nov. 15, 16, 18; III. Same title. Mitheilung aus der kais. Gesundheitsamt, Bd. **2**, 1884.

Koenig (Fr.) : "Die Tuberculose der Knochen und Gelenke, auf Grund eigener Beobachtungen," gr. 8°. 1883.

Korab : "Influence of Helenin on Bacillus Tuberculosis," Berlin med. Central-Zeitung, 1882, Nos. 30, 31.

Kredel : "Klinische Erfahrungen über Tuberkelbacillen." Bericht der Oberhessischen Gesellschaft für Natur. u. Heilkunde. Giessen, 1883.

Kussner : "Beitrag zur Impftuberculose." Deutsche med. Wochenschr, 1883. No. 36.

Lachmann : "Zur Kenntniss der Tuberkelbacilen." Dtsch. med. Wochenschr., 1884, No. 13.

Lenz, V. : "Experimentelle Untersuchungen über die Infectiossität des Blutes und Urines Tuberculöser." Dissert. Berlin, July, 1881.

Leyden (E.) : "Klinisches über den Tuberkelbacillus." Zeitschr. f. klin. Med., VIII., p. 375. — Ctbl. f. d. med. Wiss., 1885, p. 152.

Lichtheim (L.) : "Zur Diagnostischen Verwerthung der Tuberkelbacillen." Fortschritte der Medicin, Bd. 1, Hft. 1, p. 4, 1883.

Long : (staining method). Berliner klin. Wochenschr., January, 1883, p. 33 (vid. Pfeiffer, below).

Malassez (L.) et Vignal (W.) : "Sur le Micro-organisme de la Tuberculose Zooglæique."[1] Compt. Rend., T. 98, July 28, 1884, p. 203. Also 5 Nov., 1883. Soc. de Biologie, Seances, 12–19 Mai et 9 Juin, 1883. Archiv de Phys., 15 Nov., 1883.

MENSCHE (On Staining): "Vortrag, gehalten in der med. Sect. des Niedenh." Vereins für Natur. u. Heilk. zu Bonn, 1883.

MULLER (F.): "Ueber die Diagnostiche Bedeutung der Tuberkelbacillen" (Verhand'l. phys. med. Gesellsch. zu Würzburg, 1883, pp. 7, 8°.

NATHAN, J. S.: "Ueber das Vorkommen von Tuberkelbacillen bei Ottorhoen" (Aus d. med. klin. Institut zu München. Otiatrisches Ambulatorium von Dr. Bezold). Deutsches Archiv f. klin. Med. xxxv. p. 491.

NEGRI (A. F.): Coloration des spores dans bacilles de la tuberculose. Journal de Micrographie, T. viii., 1884, No. 6, p. 349.

ORTH, J.: "Notizen zur Farbetechnik." Berl. klin. Wochenschr., 1883, No. 28, p. 421.

PETERS: "Nachweis der Tuberkelbacillen in Schnitten durch die Doppel-färbung; 'Gentiana-

---

[1] "Nous avons appelé tuberculose *zooglæique* une affection causée par l'inoculation de produits tuberculeux, dans lesquels nous n'avions pas trouvé de bacilles (tubercule cutané parvi d'abces ossifluent) ayant tous les caractéres cliniques et anatomo-pathologiques de certaines tuberculoses, mais présentant, pendant les premières générations tout au moins des amas zoogloeiques de micrococques et pas de bacilles." l. c. p. 494.

violet.' Anilingelb ohne Salpetersäure entfärbung." Berl. klin. Wochenschr., 1883, No. 24, 26, p. 365.

PETRI: "Zur Färbung des Koch'schen Bacillus in Sputis sowie über das gleiche Verhalten einiger Pilzzellen." Berl. klin. Wochenschr., No. 26, p. 739, 1883.

PFEIFFER (Aug.): "Ueber die Regelmässigkeit des Vorkommens von Tuberkelbacillen im Auswurf Schwindsuchtiger." Berl. klin. Wochenschr., 1883, Jan'y, p. 32 (gives Long's method).

PFUHL-PETRI (Staining method): Deutsche militärärztliche Zeitschrift, 1884, Heft. 3.

PRIOR: "Beitrag zur Färbarkeit des Tuberkelbacillus." Berl. klin. Wochenschr., 1883, No. 33, p. 497.

PUTZ: "Ueber d. Beziehungen der Tuberculose des Menschen zur Tuberkulose des Thiere." Stuttgart, 1883.

"Photographing Bacillus tuberculosis." Journ. Roy. Micro. Soc., Ser. II., Vol. IV., 1884, pt. 4, p. 627.

QUINLAN (J. B.): "Bacillus Mounting." The Microscope, Vol. III., 1883, p. 138.—Med. and Surg. Reporter, 1883. Journal de Micrographie, 1883, p. 441.

RANSOME (A.): "Bacilli in condensed aqueous vapour of the breath of phthisical persons." Proc. Roy. Soc. XXXIV., 1882, p. 274–5.

RAYMOND ET ARNAUD: "Recherches Experimentales sur l'étiologie de la Tuberculose." Extr. des Archives Génèr. d. Méd., 1883, Jan'y.

REICHENBACH (H.): "Die Entdeckung der Tuberculose Bacillen durch Dr. Robert Koch." Humboldt, Monatsschrift für die gesammten Naturwissenschaften. Hft. VII., 1882.

REINSTADLER (F. A.): "Ueber Impftuberculose." Arch. f. exp. Pathol. u. Pharm. Bd. XI. 1879, p. 103–121.

RINDFLEISH: I. "Ueber die Methode der Bacillen färbung in Sputum." Berl. klin. Wochenschr. No. 12, 1883, p. 183. II. "Demonstration von Tuberkelbacillen." Sitzber. d. phys. med. Gesell. zu. Würzburg, Jahrgang, 1882, p. 22.

ROSENSTEIN (S.): "Vorkommen der Tuberkelbacillen im Harn." Ctbl. f. d. med. Wiss. 1883, No. 5.

RUHLE AND LICHTHEIM: "Einfluss der Entdeckung der Bacillen auf die Pathologie, Prophylaxe und Therapie der Tuberculose." Vorhandlung des Congresses fur innere Med. 1882, II. Congress.

SALOMONSEN: "Om Indpodning af Tuberkulose, särligt i Kaninens Iris." Nordiskt Medicinskt Arkiv, 1879, Bd. XV., No. 12–19.

SAUVAGE: "De la valeur diagnostique de la présence des bacilles de Koch dans les crachats." Paris (Delahaye et Lecr.), 1884.

SCHLEGTENDAL: "On Schizomycetes in Tubercular Abscesses." Fortschritte der Medicin, p. 537, 1883.

SCHMIDT (H. D.): (*Bacillus Tuberculosis.*) Louisville Med. Herald, IV. 1883, pp. 459–76, 6 figs. Considers it simply a fat crystal.

SCHILL: "Ueber den Nachweis von Tuberkelbacillen im Sputum." Deutsch. med. Wochenschr, 1883, No. 2.

SCHUCHARDT UND KRAUSE: "Ueber das Vorkommen der Tuberkelbacillen bei fungösen und scrophulosen Entzundungen." Fortschritte der Medicin., No. 9, 1883.

SENKEWITSCH (Staining method): Revue für Thierheilkunde, Bd. 7, No. 7, 1884.

SMITH (T.): "Method of demonstrating the presence of the Tubercle Bacillus in Sputum." American Monthly Micros. Journ., 1884, p. 196–9.

SPINA: "Studien über Tuberculose (Wien 1883) und deren Enwiderung durch Koch und Ehrlich." Bericht über die Sitzung des Vereins f. innere Medicin, 5 Marz, 1883.

SOMARI E BRUGNATELLI: "Studi spermentali sul bacillo della tuberculosi (Experimental studies on the Bacilli of Tuberculosis)." Redii R. Instit. Lombardo, Vol. XVI., 1883, No. 16.

STERNBERG (G. N.): "·Etiology of Tuberculosis." Abstract of a paper read before Sect. F.

(Biology) of the Amer. Assoc. Adv. Sci. Philadelphia, Sept. 9, 1884.

STOWELL (C. H.): "Bacillus staining." The Microscope, Vol. IV. 1884, No. 4, p. 79.

TAPPEINER: "Zur Frage der Constagiosität der Tuberculose, Experimentelle Untersuchungen." Deutsche Arch. f. klin. Med., Bd. XXIX., p. 595–600.

TOUSSAINT: "Sur le parasitisme de la tuberculose." Compt. Rend. T. 93 (1881), p. 350–3.

VERAGUTH (C.): "Ueber den Nachweis der Tuberkelbacillen in Chromsaure Praparaten." Berl. klin. Wochenschr. No. 13, 1883, p. 190.

WEIGERT (Staining method for tubercle bacilli): "Färbungs Methoden," by Dr. Hugo Plaut, Leipzig, 1885, p. 19.

WEICHSELBAUM (A.): "Ueber Tuberkelbacillen in Blute bei allgemeiner acuter Tuberculose." Wien med. Wochenschr., 1884, No. 12, 13. vid. also vol. for 1883. II. (Inhalation of tubercular matter). Wiener med. Jahrb. 1883.

WESNER (F.): "Ueber das Vorkommen der Tuberkel bacillen in der Organen Tuberkulöses." Deutsches Archiv. f. klin. Med. XXXIV., p. 583.

WILLIAMS (Th.): "On the relations of the Tubercle Bacillus to Phthisis." Lancet, 1883, No. 3127.

WILLIAMS (C. T.): "Influence of Culture fluids

and Medicinal reagents on the Growth and Development of Bacillus Tuberculosis." Journ. Roy. Mic. Soc., 1884, p. 932.

ZIEHL (Fr.) : I. "Zur Färbung des Tuberkelbacillus." Deutsch med. Wochenschr. 1882, No. 33, p. 451. II. Zur Lehre von der Tuberkelbacillen, inbesonderes über deren Bedeutung für Diagnose und Prognose." *Ibid.*, 1883, No. 5.

LUC (H.) : "De la tuberculose de la conjonctive compareé au lupus de cette muqueuse, contribution à la differenciation clinique de ces deux affections." 8°., 39 pp., Paris (Davy), 1883.

PFEIFFER (A.) : "Tuberkelbacillen in der Lupus erkrankten Conjunctiva." Berl. klin. Wochenschr., July, 1883, p. 431

PAGENSTECHER UND PFEIFFER : "Lupus oder Tuberculose." Berl. klin. Wochenschr. 1883, No. 19, p. 282.

SCHULLER (M.) : "Histologische Studien über Mikrokokken des Lupus." Ctbl. f. Chir., 1881, No. 46.

TYPHOID FEVER.
Bacillus typhi-abdominalis (Brautlecht).

The *Bacilli* of typhoid fever are decolorized when prepared according to Gram's general method. Klebs employed hæmatoxylin as a staining agent in his researches. There is a difference of opinion as to the specific microbe of typhoid fever being a *Bacillus*, — according to Letzerich it is a *Micrococcus*, which he demonstrated as follows: —

Letzerich's Method. — He employed partly fresh and partly hardened organs, made sections, and cleared them up in weak caustic potash solution or a solution of carbonate of soda, or, better still, in highly diluted glacial acetic acid (1 : 3), as this does not destroy the nuclei and allows the relation of the Schizomycetes to the tissue to be seen. In sections of intestine hardened in alcohol, he employed the acid one part to two of water, and allowed them to remain in it for half an hour, added a little glycerine, and mounted on a slide; this gave a beautiful preparation, in which both nuclei and microbes were apparent.

Rindfleisch's Method. — (1.) He took water from a suspected well, allowed a drop to dry upon a slide, colored it with a weak methyl-violet solution, washed it a moment in water, dried it, and mounted in Canada balsam. Result: numerous deeply blue colored rod Bacteria.

(2.) Inoculated with the water a culture of human-flesh gelatine, using all precautions against external contamination. Result: a rapid solution of gelatine by the culture.

LITERATURE.

BOENS : "La fievre typhoide, ses causes, son traitement et sa prophylaxie." Acad. roy. de med. de Belgique. Bulletin, 1883, 3d ser. T. XVII., p. 176.

BRAUTLECHT : "Typhus Bakterien und Trinkwasser." Virch. Archiv, 1880, p. 80.

EBERTH : " Neue Untersuchungen über d. Bacillus d. abdominal Typhus." Archiv f. path. Anat. u. Physiol. u. f. klin. Med. Bd. 83 (1881). II. "Die Organismen in d. Organen bei Typhus abdominalis." Virchow's Archiv Bd. 81, 1880, p. 58, vid. also Bd. 87.

FRANK (A.) : "Zur Aetiologie des abdominal Typhus." Bayr. ärtz. Int. Bl., 1881, No. 23.

FRIEDLAENDER : "Bacillus des abdominal Typhus." Sitzung d. Ver. f. innere Medecin, Berlin, 1881, 17 Nov.

GAFFKY : "Zur Aetiologie des abdominal Typhus." Mittheilung aus d. kais Reichsgesundheitsamt, Bd. 2, 1884.

HANOT (V.) : "Miliaire bactéridienne dans la fièvre typhoide." Rev. d. Med., 1881, 10 Oct.

HEIN (I.) : " Typhusbacillen im Milzblute resp. Milzsafte." Ctbl. f. d. Med. Wiss. 1884, No. 40.

KLEBS (ED.) : I. "Der Ileotyphus eine Schizomycose." Archiv f. exp. Pathol. u. Pharm. Bd.

XII. Heft. 3, p. 231 (1880). II. "Bacillus des Abdominal typhus und der typhöse Process." *Ibid.* Bd. XIII., p. 381. Taf. IV., V., VI.

KLEIN (E.) : (On Typhoid Fever). Reports of the Medical Officer of the Privy Council, 1875.

KOCH (R.) : (Typhoid Fever). Mittheilungen a. d. k. Gesundheitsamt, Bd. 1, 1881.

LETZERICH (L.) : I. "Studien über Typhus abdominalis." Virchow's Archiv, Bd. 68. II. "Experimentelle Untersuchungen über Typhus abdominalis." Archiv f. experim. Pathol. u. Pharm. Bd. 9, p. 312.

LUDWIG (E.) : "Beitrag zur Frage der Entstehung und Verbreitung des Abdominaltyphus." Würtemberger med. Corr. Bl. 1882, No. 5–6.

MEYER (WM.) : "Untersuchungen über den Bacillus des abdominal Typhus." (Aus dem städt. allg. Krankenhaus zu Berlin.) Dissert., Berlin, Sept., 1881.

RAPPIN : "Des bactéries de la bouche a l'etat normale et dans la fièvre typhoide." Paris, 1881.

RINDFLEISH : "Ueber Trinkwassertyphus." Sitzungsber. der phys. med. Gesellsch. zu Würzburg, Jahrgang 1882, p. 133.

ROTH (F.) : "Ueber die Verbreitung des Typhoides (abdominal Typhus) nach Wasserläufer." Bayr. ärztl. Int. Bl., 1881, No. 44.

TAYON : "Sur le microbe de la fièvre typhoide

de l'homme; culture et inoculations." Compt. Rend. T. 99, p. 331, Aug. 18, 1884.

TIZZONI : "Studi sulla natura del tifo abdominale." Annali universale di Medicina, Vol. 251, 1880.

WERNICH (A.) : "Typhus Bacillen." Arztl. Int. Bl. 1881, No. 44. I. " Typhus Bacillen." Sitzung des Vereins für innere Med., Berlin, 1880, 4 July. II. " Studien und Erfahrungen über den Typhus abdominalis." Zeitschr. f. klin. Med. Bd. IV., H. 1. III. " Der abdominal Typhus, Untersuchungen über sein Wesen, seine Tödlichkeit und seine Bekämpfung" 1882, Berlin (A. Hirschwald).

WHOOPING COUGH.

Burger's Method. — Cover-glass preparations made according to general rules, and stained with aqueous solutions of fuchsin and methyl-violet.

LITERATURE.

BURGER (CARL) : " Der Keuchhustenpilz." Berlin. klin. Wochenschr., Jan., 1883, p. 7.

CONCRETIONS OF THE LACHRYMAL DUCTS.
Cladothrix foersteri (Cohn).

No special methods, so far as I can ascertain, have been given for the study or preparation of the Bacteria found in the concretions of the lachrymal ducts.

LITERATURE.

Cohn (F.): Beitr. z. Biol. d. Pflanzen. Vol. 1, p. 186.

Foerster: "Pilzmasse in unteren Thränen-Kanälchen." Arch. f. Opthalm. XV., 1.

Goldzieher (W.); "Streptothrix Foersteri im unteren Thränenröhrchen." Ctbl. f. prakt. Augenheilk. 1884, Febr.

Graefe: "Ueber Leptothrix in d. Thränenröhrchen." Archiv f. Opthalm. Bd. XVI., 1.

Reuss (A. v.): "Pilzkonkretionen in den Thränenröhrchen." Wien med. Pr., 1884, No. 7 u. 8.

DENTAL CARIES.

Leptothrix buccalis (Robin) etc.

Prof. Ferd. Cohn [1] calls attention to the fact that in a letter dated September 14, 1683, A. Van Leeuwenhoek gave notice to the Royal Society that with the aid of his microscope he had discovered in the white substance adhering to his teeth very little animals moving in a very lively manner. They were the first Bacteria the human eye ever saw.

Leber's Method. — If the *Leptothrix* fibres are found in an acid medium, it is sufficient to add iodine to stain the contents blue or violet. If they

---

[1] Nature, xxix., 1883, p. 154.

occur in an alkaline medium, acidulate with dilute hydrochloric or acetic acid, then add iodine. Prepare cover-glasses according to general directions.

Miller's Method. — Dr. Miller made one thousand sections of carious teeth, of which not one failed to show Bacteria deep in the tooth tissue, whether the tooth was a living or dead one. The contrary results obtained by former observers are due, according to Miller, to their methods of coloring, which were not suited to the case in hand. To show the grouping of the Bacteria heaps, whether in transverse or longitudinal sections, use an alcoholic solution of Magdala red. Nearly as good results are given with fuchsin, methyl-blue, or Bismarck-brown.

LITERATURE.

ARNDT: " Beobacht. an Spirochaete denticola." Archiv f. path. Anat., Physiol. u. klinisch. Med., 1880, Bd. 79.

BAUME: " Odontologische Forschungen." II.

LEBER UND ROTTENSTEIN: " Untersuchungen über Caries der Zähne." Berlin, 1867.

MILLER (W. D.): I. " Der Einfluss der Microorganismen auf die Caries der menschlicher Zähne." Archiv f. experim. Pathol. u. Pharm. XVI., 1882. II. " Ueber der Caries der Zähne." Corresp. f. Zahnärzte. Bd. XIII. III. " Ueber einen Zähnspaltpilze, Leptothrix gigantæ." Ber d. deutsch.

bot. Gesellsch. 1883, H. 5. IV. "Zur Kenntniss der Bakterien in der Mundhöhle." Deutsche med. Wochenschr. 27 Nov. 1884. No. 48, p. 781.

AREA CELSII.
(Alopecia areata.)

Buchner's Method was to remove a diseased hair with sterilized pincers, and place it entire into ordinary culture fluids.

Von Sehlen's Method for staining the micrococci of this disease is to place the hairs *in toto*, after having removed all oil and fat by means of chloroform and ether, in a very dilute anilin-oil-fuchsin solution, or a carbolic fuchsin solution for twenty-four hours, then wash in hydrochloric acid alcohol, and remove the acid with distilled water. Give double-staining with methyl-blue or gentian-violet solution, wash in absolute alcohol, pass through oil of cloves, and mount in Canada balsam. Simple staining with one aniline color leads to no result, because the hair cells color in a similar manner. *For transverse sections*, dehydrate the hairs with absolute alcohol, and allow them to lie for some time in chloroform, then imbed in a chloroform paraffine solution, and cut sections 0.01. mm. thick, fasten these to the slide by the ordinary methods, *e. g.*, by painting the slide over with a mixture of oil of cloves and collodion before placing the sections upon it, and allowing it to dry before proceeding further) and stain with a strong solution of aniline-oil-fuchsin, or of carbolic fuchsin, then wash in acidulated (HCl) alcohol, and remove the acid with distilled water. Give second staining with a concentrated aqueous solution of gentian-violet, wash in absolute alcohol, pass through oil of cloves, and mount in Canada balsam.

LITERATURE.

BALZER : "Contribution à l'etude de l'erythème tricophytique." Archiv de Physiol. 3° sér. 1883, T. 1, p. 171.

BUCHNER (H.) : "Kritische Bemerkungen zur Aetiologie der Area Celsii." Virchow's Archiv, Bd. 74. p. 527. (1878).

MICHELSON (P.) : "Bemerkung zu den Arbeiten des Herrn Dr. v. Sehlen über die Aetiologie der Alopecia areata (area celsii)." Virchow's Archiv, Bd. 99, Hft. 3, p. 572.

VON SEHLEN (D.): "Mikrokokken bei Area Celsii." Fortschritte der Med. Bd. 1. No. 23. p. 763. (1883). II. "Zur Aetiologie der Alopecia areata." Virchow's Archiv, Bd. 99, 2 Hft. (1885).

CHICKEN CHOLERA.
(*Micrococcus gallicidus.*)

There is some uncertainty as to whether the microbe of this disease is a dumb-bell *Mi-*

crococcus (*Diplococcus*), or a *Bacterium*. Klein intimates that Pasteur used impure cultures, and that the organism is probably a *Bacterium termo*.

Barthelemy's Method. — A hen died of this disease after having laid fourteen eggs. Barthelemy incubated these eggs, marking them in order to distinguish them from a second batch used as a 'control' experiment. Underneath the shell and on the surface of the allantois he found lacunæ of blood, black, and having a special odor like that of fowls dead of the cholera. This blood was filled with Bacteria, while the amniotic fluid contained very minute monads. It is evident the egg contained the germs of the microbes with which the liquids of the mother were filled, but these did not develop until the allantois furnished oxygen.

Pasteur's Method. — Neutralized urine and yeast water were first tried as culture fluids, but found unadapted to the purpose. A meat or chicken broth neutralized with carbonate of potash and sterilized at a temperature of $100°–115°$ C., was found to be well suited; the Bacteria developing rapidly within one hour. After cultivating the microbes for several days he filtered them all out and inoculated a chicken with the filtrate, producing the disease in a mild form.

LITERATURE.

BABES: (On chicken cholera). Archiv de Physiol. July, 1883, p. 49.

BARTHELEMY (A.): "De l'incubation des œufs d'une poule atteinte du cholera des poules." Compt. Rend. T. 96. p. 1322.

PASTEUR (L.): I. "Verbal Observations on Chicken Cholera." Compt. Rend. T. 79-89. II. "Behavior of Bacteria of Chicken Cholera to cold." Ibid. 79. No. 24. III. "Sur les maladies virulente et en particulier sur la maladie appelée vulgairement choléra des poules." Ibid. T. 90. p. 239. (1880). German translation in Arch. f. experim. Pathol. Bd. 12, Heft. 4, p. 344. IV. "Sur le Choléra des poules; études des conditions de la non-récidive de la maladie et de quelques autres de ses caracters." Compt. Rend. T. 90 (1880). p. 952 and 1030. Compare also Transactions of the International Medical Congress in London, 1881, vol. I., p. 87.

PERRONCITO: "Ueber das epizootische Typhoide der Hühner." Archiv f. wissenschaftl. u pract. Thierheilkunde, 1879, p. 22.

SEMMER: "Huhnerpest." Dtsch. Zeitschrft. f. thier. Med. u. vergl. Pathol., 1878.

TOUSSAINT (H.): "Identite de la septicémie expérimentale aiguë et du cholera des poules." Compt. Rend. T. 91, p. 301 (1880).

DIPHTHERIA.

Micrococci diphtheriticus (Cohn.)

Both *Micrococci* and *Bacilli* occur in diptheritic membranes, and there is a division among investigators as to which is the *vera causa morbi*. Buhl, Huter, Formad, Klein, and others claim *Micrococci* to be the active agents; while, as will be seen below, Loeffler considered these to be of only secondary importance. Further investigations should be made with pure cultures and inoculations from these.

Loeffler's Method. — On account of the great variety of Bacteria inhabiting the mucous membranes of men, Loeffler desired a staining fluid by which all known Bacteria might be stained, and used as such the following: —

Add to 30 cc. of concentrated alcoholic methyl-blue solution 100 cc. of caustic potash solution (1 part to 1,000 parts of water). It is sufficient for sections to allow them to remain only a few moments in the solution, which will give an intense stain to all known Bacteria. After staining, the sections are placed for a short time in a half per cent acetic acid solution, moved about for several seconds, and then dehydrated in absolute alcohol, passed through oil of cedar, and mounted in Canada balsam. By this process Loeffler found two peculiar microbes constantly present in diphtheria:

(1) A chain-building *Micrococcus*, which he cultivated according to general rules, in stiff blood-serum and on cooked potato, at ordinary temperatures and in a breeding oven. (2) A *Bacillus* like that discovered by Klebs ("für bacillar Diphtheria"); this flourished best at 37° C. on a mixture of three parts of calves' or sheep's serum to one part of neutralized veal broth, to which one per cent peptone, one per cent beet sugar, and one per cent salt is added. These *Bacilli* Loeffler regarded as important in the pathogenesis of diphtheria, while the micrococci he considered as of only secondary importance.

LITERATURE.

BIRSCH-HIRSCHFELD: "Archiv für Heilkunde." 1872, p. 389.

BUHL: "Micrococci of Diphtheria. Zeitschrift für Biologie," 1867, III.

EBERTH: Zur Kenntniss d. Bact. Mykosen, 1872.

FORSTER: "Miasmatische Verbreitung d. Diphtherie." Wien. med. Wochenschr. 1881, No. 24.

GERHARDT UND KLEBS: " Diphtherie, ihre parasitäre Natur, Verhälteniss des localen Prozesses zur allgemeinen Inflection, Contagiosität, Therapie (Chirurgie) und Prophylaxe." Verhandlung des Congresses für innere Medicin, 1882, II. Congress.

Hoffmann: "Ueber Bakterien." Botan. Zeitung, 1869.

Hüter u. Tommasi: (Micrococci of Diphtheria). Med. Centralblt., 1868, p. 177, 531, 547.

Klebs: Archiv f. experiment. Pathol. u. Pharm. IV.

Letzerich: " Diphtherie-Pilze." Virchow's Archiv, 1872. II: " Untersuchung über die morphologischen Unterschiede einiger pathogenen Schizomyceten." Archiv f. exper. Path. Bd. XII., H. 5, p. 351 (1880).

Limmer: " Ueberträgung der Diphtherie durch Hühner." Aerztl. Int. Bl., 1881, No. 31.

Loeffler: "Untersuchungen über die Bedeutung der Mikroorganismen für die Entstehung der Diphtheritis beim Menschen, bei Taube und beim Kalbe." Mittheilung a. d. kais. Reichsgesundheitsamt, Bd. 2, 1884.

Nassiloff: "Ueber die Diphtheritis." Virchow's Archiv, vol. 50, p. 350.

Oertel: " Experimentelle Untersuchungen über Diphtherie." Deutsche Archiv f. klin. Med. VIII., 1871.

Talamon (Ch.): "Note sur le microbe de la diphtherie." Progres méd. 1881, No. 7.

Wood and Formad: "The nature of the poison of diphtheria." Med. Times and Gaz., 1880, Dec. 4; also Bull. National Board of Health, No. 17, 1882.

ERYSIPELAS.

(*Micrococci.*)

Fehleisen's Method.— Material for making pure cultures was obtained from the contents of a freshly opened erysipelas bulla, and also by cutting out a bit of the skin of the affected part, having first cleaned it with ether and then with a corrosive sublimate solution. These he placed in nutritive gelatine and stiffened blood serum, and cultivated according to general rules (q. v.). With the pure cultures thus obtained he vaccinated, among others, a woman (58 years old) who was about dying with multiple sarcoma of the skin, and produced a typical case of erysipelas.

These inoculations upon human beings were justifiable, because they were undertaken with a view to cure certain tumors. Thus one case of lupus, one case of cancer, one case of sarcoma, were considerably affected and to the good of the patient. Fehleisen also, in several instances, succeeded in second inoculations, after the lapse of a few months. The vitality of this microbe is destroyed by three per cent solutions of carbolic acid, or one per cent solutions of corrosive sublimate.

LITERATURE.

BÄÄDER: "Zur Aetiologie des Erysipels," Schweis. naturf. Gesellsch. Basel, 1875, p. 314.

EHRLICH: "Ueber Erysipelas." Langenbeck's Archiv, Bd. 20, p. 418.

FEHLEISEN: I., "Ueber die Züchtung der Erysipelkokken auf kunstlichem Nährboden und ihre Uebertragbarkeit auf den Menschen." Sitzungsber. d. phys. med. Gesellsch zu Würzburg, Jahrgang, 1883, No. 1, p. 9; *ibid.* No. 8, Jahrgang, 1881, p. 126–128. II., "Die Aetiologie des Erysipelas." Mit. 1 lith. Tafl., Berlin (Fischer), 1883.

LUKOMSKY (W.): "Untersuchung über Erysipel." Virchow's Archiv, vol. 60, p. 418.

ORTH: "Untersuchungen über Erysipel." Arch. f. experim. Pathol., Bd. I., Hft. III.

RECKLINGHAUSEN UND SANKOWSKY: "Ueber Erysipel." Virchow's Archiv, Bd. 60, p. 418.

TILLMANS: "Ueber Erysipelas." Verhandlung d. dtsch. Gesellsch. f. Chir., 1878, p. 211.

WOLFF: "Ueber Erysipelas." Virchow's Archiv, Bd. 81, p. 173.

FURONCLE.

(*Micrococci.*)

According to Pasteur's description these microbes have a *Sarcina*-like arrangement. "Couples de deux et quatre grains et paquet de ces memes grains." They develop within less than six hours after being put into the culture fluid. He considers them to be the same microbes as those producing acute infectious osteomyelitis, and regards the latter disease as a furoncle of the marrow of the bone.

Pasteur's Method.—Pasteur removed from furoncles small amounts of pus, which he cultivated in chicken broth or in an infusion of yeast ("l'eau de levûre") at 35° C. He inoculated rabbits with these pure cultures, and produced boils. This microbe does not thrive in the blood. Stain by general methods.

LITERATURE.

GRAF (FR.) : " Die Anticepsis in der Ohrenheilkunde." Berlin. klin. Wochenschr. 1883, No. 14, p. 209.

LOWENBERG (B.): I., "Untersuchungen über Auftreten und Bedeutung von Coccobacterien bei eitrigen Ohrenflusse und über die durch Gegenwart bedingten therapeutischen Indicationen." Zeitschr. f. Ohrenheilkunde, X., 1882. II., "Recherches sur la presence de micrococcus dans l'oreille malade, considerations sur le rôle des microbes dans la furoncle auriculaire et le furonculose générale, applications therapeutique." Compt. Rend. T. 91, p. 555, 1880.

OGSTON: "Ueber Abscesse." Arch. f. klin. Chir., Bd. 25.

PASTEUR: "Germes in Furoncles." Compt. Rend., T. 90, p. 1036.

GONORRHŒA.

(*Micrococcus gonorrhœa.*)

Neisser's Method.— *Staining.* (*a*) Spread out the secretion in the thinnest possible layer, upon a very thin sterilized cover-glass; this is done best by allowing a small drop to flow between two cover-glasses, after which they are drawn apart. (*b*) Allow the preparation to dry in the air, heating it slowly up to 120° or 105° C., at which temperature allow it to remain for one or two hours. (*c*) Preparations must be well stained. All the basic aniline dyes may be used, and also the same bases united with an acid, as methyl-violet, gentian-violet, fuchsin, Bismarck-brown, etc., but none are so good as methyl-blue, which shows a great affinity for " cocci." Allow the prepared cover-glasses to lie for from one-half to twenty-four hours in a concentrated aqueous solution of the dye stuff. Wash in absolute alcohol, pass through oil of cloves, and mount in Canada balsam. Study with a 1-12th homogeneous immersion lens and an Abbe's condenser.

For cultures use a neutralized peptone beef-extract-gelatine, a blood-serum, or a peptone-gelatine. In the deep parts of the culture medium, where the air did not penetrate, the growth was slow, as was also the case when the gelatine was very firm.

Inoculation. Dogs have immunity from this disease, inoculations producing a simple balanitis; this is likewise the case with rabbits. A man forty-six years old, into whose urethra some of the pure culture was injected with a sterilized syringe, had, as a consequence, a true gonorrhœa. Inoculations were made upon apes, dogs, cats, and rabbits, on the conjunctiva, cornea, and urethra, by Neisser, Leistikow, Krause, and others.

LITERATURE.

AUFRECHT: " Mikrokokken in den inneren Organen bei Nabelrenen Entzündung Neugeborener." Ctbl. f. d. med. Wiss., 1883. No. 16, p. 273.

BEAUVAIS: " De la Balanite." Gaz. des Hospitaux, 1874, p. 867 e 876.

BOCKHART: " Beiträge zur Aetiol. u. Pathol. des Harnröhren Trippers." Würzb. phys. med. Ges., Jahrgang, 1883. No. 1, p. 13. No. 2, p. 17.

BOKAI (A.): " Ueber das Contagium der acuten Blennorrhœa." Centralbl. f. d. med. Wiss. 1880. 74.

ESCHBAUM: "Ein Beitrag zur Aetologie der gonorrhoischen Secrete." Deutsch. med. Wochenschr. 1883, No. 13.

FINKELSTEIN: Prager med. chir. Presse, May, 1880.

HAAB: "Der Mikrokokkus der Blenorrh. neanatorum." Festschrift. Weisbaden, 1881.

HIRSCHBERG UND KRAUSE: "Untersuchungen über Tripperbacterien zur pathologie der ansteckenden Augenkrankheiten." Centralbl. f. pract. Augenheilkünde. Sept. 1881. Berl. klin. Wochenschr. 1882. No. 32, p. 500.

KRAUSE (F.). "Die Micrococcen der Blennorrhœa neanotorum." Centralbl. f. pract. Augenheilkunde. Mai, 1882.

LEISTIKOW. "Untersuchungen über Tripperbacterien." Gesellsch. d. charité Aerzte in Berlin, 16 Feb. 1882; also, Berl. klin. Wochenschr. 1882. No. 32, p. 500; also, No. 47, p. 719. II. "Ueber Bakterien bei den venerischen Krankheiten." Charité Ann. VII. (1882), p. 750.

NEISSER (A.): "Die Mikrokokken der Gonorrhoe." Dtsch. med. Wochenschr. 1882, p. 279–282; also Med. Centralbl. 1879, No. 28.

RETER: "Ueber eine der Gonorrhoe eigentümliche Mikrokokkenform." Centralbl. f. d. med. Wissensch. 1879, No. 28.

RUCKER (G.): "Ueber Polyarthritis gonorrhoica." Dissert. in Berlin, 1880.

WEIS (F.): "Le microbe du pus blennorrhagique." Thèse de Nancy, 1880; also Annale d. Dermatol. 1881. Heft. I.

WELANDER (E.): "Quelques recherches sur les microbes pathogénes de la blenorrhagie." Gaz. med. de Paris, 1884, No. 23.

OSTEOMYELITIS.
Micrococci.

Becker was the first to make pure cultures of this microbe, which Pasteur regards as identical with that found in furoncles.

Krause's Methods; Cultures. Remove pus with antiseptic precautions and place in a sterilized test tube, closing the same with cotton. Use for culture media sterilized coagulated sheep's-blood serum or beef-juice-peptone-gelatine, or beef-juice agar agar. In both the latter mixtures use peptone siccum, 1-2 per cent; sodium chloride, 0.5-1 per cent; sodium phos. sufficient to neutralize, gelatine, 5 per cent; or agar agar, 1-1.5 per cent. Keep object glasses bearing cultures in a moist chamber and in the breeding oven. The smell of the cultures of this microbe is characteristic (" verdorbenen Kleister ") like spoiled paste, and is apparent the moment the lid of the moist chamber is raised. Cultivate also on slices of boiled potato, and notice the effect on milk which is rendered sour. Sterilize the milk in a water-bath at 100° C. and inoculate it with some pure culture; it becomes lobbered and has a sour reaction.

Staining. Prepare and stain according to Gram's general method. If it is desired to give a double staining, use a thin solution of vesuvin.

Pasteur considers this microbe identical with that of furoncles, and regards osteomyelitis as a furoncle of the marrow of the bone.

LITERATURE.

BECKER: "Osteomyelitiskokken." Verein f. innere med. zu Berlin. Dec. 17, 1883; also, Deutsch. med. Wochenschr. 1883. No. 46 and No. 52, p. 816. II.: Mikrokokken der Osteomyelitis. Berl. klin. Wochenschr, 1883. No. 52, p. 816.

EBERTH: "Primäre infectiose Periostitis." Virchow's Archiv, Bd. 65, p. 352.

FEHLEISEN: (Panaritium) Phys. med. Gesellschaft. Würzburg. 1882, p. 113–121.

KOEHER: "Die acute Osteomyelitis." Dtsch. Zeitschrift f. Chir. Bd. XI., p. 96.

KRAUSE (F.): "Ueber einen bei der acuten infectiösen Osteomyelitis des Menschen vorkommenden Mikrokokkus." Fortschritte der Medicin, Bd. 2. No. 8. 1884. p. 261.

LÜCKE: "Die primäre infectiose Knochenmark und Knochenhautentzündung." Dtsch. Zeitschrift. f. Chir. Bd. IV., p. 231.

PASTEUR: Compt. Rend. T. 90, p. 1036.

RODET (A.): "Etude expérimentale sur l'osteomyélite infectieuse." Compt. Rend. T. 99. (Oct. 6, 1884) p. 569.

ROSENBACH: "Beiträge zur Kenntniss der Osteomyelitis." Dtsche. Zeitschrift f. Chir. Bd. X., p. 382.

SCHÜLLER: "Zur Kenntniss der Mikrokokken bei acuter infectioser Osteomyelitis, Mikrokokken heerde im Gelenkknorpel." Ctbl. f. Chir. 1881. No. 42.

ACUTE CROUPOUS PNEUMONIA.
Micrococci pneumoniae infect.

Friedländer's Method. Cultures. From the lung of a patient who has died of this disease take small pieces, by means of a sterilized wire, and cultivate in blood-serum according to Koch's plan. In taking the material it must first be seen to that the lungs are perfectly fresh, *i. e.*, free from all appearances of decomposition; and secondly, that contamination from without is avoided. Cut pieces from the pleural surfaces, or with a heated knife cut out a piece of lung, and cut successive slices in different directions from this until a piece is obtained which is completely free from external contamination. The blood serum being inoculated in test tubes closed with sterilized cotton, is placed in a thermostat at the temperature of the body. Instead of blood serum Dr. Frobenius used an infusion of meat, peptone, and salt. Numerous generations should be made to ensure pure cultures.

The pure cultures have a characteristic "nail"

shape in gelatine. Very good cultures in grayish drops may be made on slices of boiled potato, — these give the characteristic "nail" shape when inoculated back on to gelatine.

Inoculation. Remove a bit of a pure culture by means of a sterilized needle, and place it in a watch-glass previously sterilized, and holding some sterilized distilled water; by mixing this a milky fluid is obtained, which is drawn up into a previously heated Pravaz syringe, and after cleaning the skin of the animal with a one per cent corrosive sublimate solution, is injected into the lung, or the animal is made to breathe the spray of the same fluid produced with an atomizer properly sterilized, and operated through a long rubber tube, in order that the investigator may not inoculate himself.

Staining. Pneumonia *Micrococci* are characterized in many cases by a peculiar capsule. Water and alkalies destroy this, and Fränkel and Sée even assert that the capsule does not belong to pneumonia *Micrococci*, but is a result of the mode of preparation and staining. Gentian-violet and fuchsin stain the capsule only slightly, while giving the *Micrococci* a deep color. Bismarck-brown and methyl-blue color the capsule and coccus of about alike intensity. Salvioli and Zaslein obtained the best preparations by staining with a mixture of Bismarck-brown and methyl-violet.

To render the Micrococci visible in tissues, allow the sections cut from the hardened material to remain for one hour in aniline-gentian-violet solution [F. 19.] at 45° C. Wash in iodo-iodide of potash solution [F. 35 d], pass through absolute alcohol, oil of cloves, and mount in Canada balsam. At times the stained *Micrococci* give up their color when treated with alcohol after the potash solution.

To render the capsules visible in exudates: (1) prepare cover-glasses according to general rules, lay in acetic acid, color in gentian violet or Bismarck brown; pass through alcohol, oil of cloves, and mount in Canada balsam; or (2), allow the staining fluid to act for twenty-four hours; or (3), allow the cover-glasses to float in aniline water-gentian-violet, and warm to steaming for one minute, then wash one half minute in alcohol, and study in distilled water; or (4), use a weak eosin solution for twenty-four hours; or (5), osmic acid, which causes the contours of the capsules to stand out sharply.

LITERATURE.

BRUYLANTS ET VERRIERS: Bull. de l'Acad. Belg. 1880.

EBERTH: Dtsch. Arch. f. klin. Med. Bd. 28.

FRIEDLÄNDER (C.) : I. " Die Mikrokokken der Pneumonie." Fortschritte der Medicin, Bd. 1,

1883. No. 22, p. 716. II. "Weitere Bemerkungen über pneumonic Micrococcen" *Ibid.*, Bd. 2, No. 10, p. 340, 1884. III. "Ueber die Schizomyceten bei der acuten fibrinösen Pneumonia." Virchow's Archiv, Bd. 87, p. 319. IV. "Den Micrococcus der croupösen Pneumonie." Berliner klin. Wochenschr. 1883, No. 48, p. 752.

FRIEDLÄNDER U. FROBENIUS : "Der Micrococcus der croupösen Pneumonie." Berl. klin. Wochenschr., 1883. No. 48, p. 752.

GILES : British Med. Journ., July 7, 1883.

GRIFFINI U. CAMBRIA : "Croupöse Pneumonie." Ctbl. f. d. med. Wiss. 1883, p. 334.

GÜNTHER U. LEYDEN (First description of) : Sitzungsbericht des Vereins für innere Med., Nov. 20, 1882.

JURGENSEN : "Pneumonie Micrococci" (Congress fur innere Med.). Fortschritte der Medicin, Bd. 2, Heft. 9, p. 333 (1884).

KLEBS : Arch. f. experim. Pathol., Bd. IV.

KLEIN (E.) : " Ein Beitrag zur Kenntniss der Pneumokokkus." Ctbl. f. d. med. Wiss. 1884, No. 30.

KOCH (Acute phenomena following Recurrent fever) : Mittheilung. a. d. kais. Gesundheitsamt., Berlin. Bd. 1, 1881.

LETZERICH : Morphologische Unterschiede einiger pathogenen Schizomyceten." Archiv f. experim. Pathol., Bd. VII, H. 5, p. 351 (1880).

Mátroy: "Pneumoniekokken." Wiener med. Presse. 1883. No. 23, 24.

Mendelson: "Pneumoniekokken." Zeitschr. f. klin. Med. 1883, VII., p. 206.

Poels (J.) u. Nolen (W.): "Die Mikrokokken der Pneumonie des Menschen und der Lungenseuche der Rinder." Ctbl. f. d. med. Weiss. 1884, No. 9.

Purjesz (S.): "Die Aetiologie der crouposen (fibrinösen) Pneumonie." Deutsch. Arch. f. klin. Med. XXXV., p. 301.

Salvioli u. Zaslein: "Ueber den Mikrokokkus und die Pathogenese den croupousen Pneumonie." Ctbl. f. d. med. Wiss. 1883, No. 41, p. 721.

See (G.): "Sur les pneumonies infectieuse et parasitaires." Comp. Rend., Nov. 24, 1884. T. 99, p. 931.

Strassmann (F.): (Micrococci of Pneumonia). Stzb Jenaisch. Gesell. f. Naturwiss. 1883, 1884, p. 16, 17.

Vierordt (O.): "Zur Kenntniss des Vorkommens von Spiralenbildung im bronchial Secret." Berl. klin. Wochenschr. 1883, No. 29.

Ziehl (F.): "Ueber das Vorkommen der Pneumoniekokken im pneumatischen Sputum." Ctbl. f. d. med. Wiss. 1883, No. 25, p. 433. II. "Ueber den Nachweis der Pneumoniekokken im Sputum." Ibid. 1884, No. 7.

RECURRENT FEVER.

Spirochæte obermeyeri (Cohn).

In most cases the microbes of this disease cannot be studied in the tissues, since the *Spirilla* seem to be differently constituted from other Bacteria, being entirely destroyed by acids, alkalies, and even by distilled water. They act much more like protoplasm than like nuclei, consequently they are not stained by the usual methods of staining nuclei, as are most other Bacteria. Koch is the only one who has stained them (with brown aniline), and photographed them, and he declares the identification of *Spirilla* in hardened organs to be a very different matter.

Koch's Method for cover-glass preparations is to dry the blood upon the slide, and stain with fuchsin, methyl violet, or gentian violet, or methyl blue.

Friedländer's Method of obtaining the microbes is to take some blood from the patient by means of the cupping-glass, and allow it to coagulate. The *Spirilla* gradually gather upon the surface of the clot, often in large groups of twenty or more twisted up in a glomerule. They will retain their vibratile movements for hours, even for days, outside of the body, and it has been recently shown (Albrecht) that they also increase outside the organism. They cause, by their combined movements, strong currents, so that they may be examined with quite low powers. They are only to be found in the blood during febrile periods.

LITERATURE.

ALBRECHT: I. "Zur Entwickelung der Spirochaete obermeyeri." Archiv klin. Med., Bd. 19, 1. II. "Eine einfache Method zur mikroskopische Untersuchung des Blutes auf Spirillen." Med. chir. Rundschau, Bd. XIX. s. 508.

BIRSCH-HIRSCHFELD: "Ueber die Recurrens-Spirochaeten." Med. Jahrb. Bd. 166, Heft. 2, p. 211.

CARTER: (Production of Relapsing fever in monkeys, by the inoculation of human blood containing Spirillum Obermeyeri). Lancet. 1879, vol. 1, p. 84. 1880, vol. 1, p. 662.

COHN (F.): "Zur weiteren Kenntniss des Febris recurrens und der Spirochæten." Deutsche med. Wochenschr., 1879, No. 19.

ENGEL: "Ueber d. obermeyer'sch Recurrens Spirillen." Berl. klin. Wochenschr., 1873, p. 409.

HEYDENREICH: "Untersuchungen über d. Parasiten d. Ruckfalltyphus." Berlin: 1877; also, St. Petersburger Wochenblatt, 1876.

KOCH (R.): Deutsch. med. Wochenschr. No. 19, 1879.

MANASSEIN: St. Petersburger Wochenblt., 1876, No. 18.

MOLCHUTCHOWSKY: *Ibid*, 1876. No. 6.

OBERMEIER: "Vorkommen feinster, eigene Bewegung zeigender Fäden im Blut von Recurrens-Kranken." Med. Centralbl. Bd. XI, 10, 1873. Berl. med. Gaz. Marz, 1873; Berl. klin. Wochenschr., 1873, p. 152 and 391.

WEIGERT: "Bemerkungen über die obermeier'schen Recurrens-faden." Dtsch. med. Wochenschr., 1876, p. 471–498.

YELLOW FEVER.

Micrococci (*Cryptococcus*).

Friere's ethod was to cultivate the microbes obtained from the blood of yellow-fever patients according to the general methods, and to attenuate the cultures by heat in a manner similar to that employed by Pasteur for anthrax. In November, 1883, he received permission from the Emperor of Brazil to vaccinate human beings; this he did to the extent of several hundreds, giving them the disease in a benign form. It is not known how long the immunity conferred by inoculation lasts, but at first it is absolute.

LITERATURE.

FRIERE (D.) ET REBOUGEON: "Le microbe de la fievre jaune. Inoculation, Preventive." Compt. Rend. de l'Acad. de Sc. de Paris, T. 99, Nov. 10, 1884, p. 804.

LECAILLE: "Le microbe de la fièvre jaune." Journal de Micrographie. T. 8, p. 75.

MISCELLANEOUS PATHOGENIC BACTERIA.

Not finding any special methods given for the study of the following forms, which can be studied by the general rules given in Chapter I, or by any of the numerous modifications of these rules given for the various "specific" microbes, I will simply enumerate them, and give their literature, that further research in regard to them may be incited and facilitated.

CONTAGIOUS SEPTICÆMIA.
Micrococcus septicus (COHN).

Besides the septicæmia produced by *M. septicus* (Microsporon septicum Klebs), Sternberg claims to have produced the disease in rabbits by the *Micrococci* obtained from his own saliva, while Koch produced a septicæmia in mice by the injection of a putrescent meat broth under the skin of the back. Extensive gangrene, with much œdematous exudation, followed, and death ensued in two days and a half. The blood, the capillaries of the kidney, and the enlarged spleen, contained numerous oval *Micrococci*, singly as dumb-bells, and in zoogloea.

LITERATURE.

ARDENNE: "Les Microbes, les Miasmes, et les Septicæmies." Paris, 1882.

BILLROTH: "Untersuchungen über die Vegetationsformen der Coccobacteria Septica." Berlin, 1874, pp. 200.

BINZ: "Ueber den Einfluss innere Arzmittle auf die Septicæmieund andere Infectionskrankheiten." Wien, Med. Pr., 1881, No. 38.

DEGAGNY (C.): "Sur le micro-organisme d'une septicémie observée chez l'homme et le mouton." Soc. de Biol. 17 Mai, 1884.—Journal de Micrographie, 1884, p. 348.

GAFFKY: "Experiment Erz. Septicæmia." Mittheilung. a. d. kais. Gesundheitsamt, Bd. 1., 1881, Berlin.

KOCH (R.): "Untersuchungen über die Aetiologie d. Wundinfections-Krankheiten." Leipzig, 1878.

LITTEN: "Septicæmie." Danziger Naturf. Vers., 1881.

ROSENBERGER: I. "Ueber Septicæmie." Phys. med. Gesellsch. zu Würzburg, 1882, p. 41, 45. II. "Experimentelle Studien über Septicæmie." Ctbl. f. d. med. Wiss, 1883, No. 4.

STERNBERG, (G. M.): "Fatal Form of Septicæmia in the Rabbit, produced by the Subcutaneous injection of Human Saliva." Report,

Nat. Board of Health Bull. April 30, 1881, 22 pl., 1 pl.

TIEGEL: "Ueber die fiebererregende Eigenschaft des Microsporon septicum." Dissertation, Bern. 1871. See also, Correspondenzblatt für Schweizer Aerzte 1871, p. 275.

WAGNER (P.) : "Ueber Aetiologie und Symptomatologie der kryptogenetischen Septicopyæmia." D. Archiv f. klin. Med. Bd. XXVIII, p. 562.

RHEUMATIC ARTHRITIS.
Micrococcus rheumarthritis.

LITERATURE.

LEYDEN (M. H.): "Gelenkenrheumatismus Micro-organismen (*Streptococcus*)," Dtsch. med. Wochenschr., 1882, p. 656.

KUSSMAUL: "Mikrokokken bei Gelenkrheumatismus," Roser's und Wunderlich's Archiv f. physiol. Heilkunde, Bd. XI., p. 626.

ENDOCARDITIS ULCEROSA.
LITERATURE.

EBERTH (C. J.): " Ueber diptheritische Endocarditis." Virchow's Archiv, vol. 57, p. 228.

HEIBERG: Ein Fall von Endocarditis ulcerosa puerperalis, mit Pilzbildung im Herzem (Mycosis endocardii). Virchow's Archiv, vol. 56, p. 407.

KLEBS: "Weitere Beitrag zur Entstehungsgeschichte der Endocarditis," Archiv f. experim. Pathol., Bd. IX., 1873, p. 52.

KOESTER: "Die embolische Endocarditis." Virchow's Archiv, vol. 72, p. 257.

MAIER (R.): "Ein Fall von primärer Endocarditis diptheritica." Virchow's Archiv, vol. 62, p. 145.

INFECTIOUS WOUND DISEASES.
Micrococci.
LITERATURE.

KOCH (R.): "Ueber d. Aetiologie d. Wundinfections-krankheiten," Leipzig, 1878, Ctbl. f. med., Wiss., 1879, p. 175.

ROSENBACH (F. J.): I. "Micro-organismen bei den Wundinfectionskrankheiten der Menschen," Wiesbaden (J. F. Bergmann), 1884. II. "Untersuchungen über die kleinster, lebender Wesen zu den Wundinfectionskrankheiten der Menschen," Wiesbaden, 1885; III. "Ueber einige fundamentale Fragen in der Lehre von der chirurgischen Wundinfectionskrankheiten. Giebt es verschiedene Arten der Faulniss?" Dtsch. Zeitschr. f. Chir., Bd. XVI., p. 342–368.

VERNEUIL (A.): "De l'auto-inoculation traumatique," Revue de Chir., Bd. III., No. 12, Dec. 12, 1883, p. 921–952.

WASSILIEF (N. P.): "Beiträge zur Frage über die Bedingungen unter denen es zur Entwickelung von Mikrokokken Colonien in den Blutgefassen kommt." (Aus dem Strasburger pathol. Inst.) Ctbl. f. d. med. Wiss., 1881, No. 52.

WOLFF : " Zur Bacterien Lehre bei accidentellen Wundkrankheiten," Virchow's Archiv f. pathol. Anat. u. Physiol., Bd. 81, p. 139.

VARIOLA.
Micrococcus variolæ et vaccinæ (Cohn).

To positively connect the *Micrococci* in this case with the disease is a matter still unaccomplished. A series of accurate and careful culture experiments should be made, absolutely pure cultures obtained, and re-inoculations made to ascertain whether the disease can be reproduced. To stimulate such a research, the Grocers' Company of London have offered a prize of £1000.[1]

---

[1] The Worshipful Company of Grocers' First Quadrennial Discovery-Prize of £1000, 1883-1886, in aid of original research in sanitary science, have announced the following problem: —

"To discover a method by which the vaccine contagion may be cultivated apart from the animal body, in some medium or media not otherwise zymotic — the method to be such that the contagion may, by means of it, be multiplied to an indefinite extent in successive generations, and that the product after any number of such generations shall (so far as can within the time be tested) prove itself of identical potency with standard vaccine lymph."

The prize is open to universal competition, British and foreign. Competitors for the prize must submit their respective treatises on or before the 31st of December, 1886; and the award will be made as soon afterward as the circumstances of the competition shall permit — not later than the month of May, 1887. In relation to the Discovery-Prize, as in relation to other parts of the company's scheme in aid of sanitary science, the court acts with the

LITERATURE.

CARPENTER: (Vaccine). vid. "Paper on the Germ Theory of Disease from a Natural History Point of View." British Assoc., 1883.

CHAUVEAU: "Nature du virus-vaccin, Détermination expérimentale des elements qui constituent le principe actif de la serosité vaccinale virulente." Compt. Rend. T. 66 (1868), p. 289, 317, 359, 948. Chauveau was the first to prove experimentally that the active principle of the virus of vaccinia and variola is a particulate nondiffusible substance.

COHN: "Micrococcus vaccinæ," Virchow's Archiv, 1872.

CORNIL ET BABES: "Note sur le siège des bactéries dans le variole, la vaccine et l'erysipèle," 8°, 6 pp., Paris (Alean-Levy), 1884, Extr. de l'Union med. VIII., 1883.

JOLYET: "Sur l'etiologie et la pathologénie de la variole du pigeon, et sur le developpement des microbes infectieux dans lymph." Compt. Rend. T. 92, p. 1522.

KLEBS (E.): I. "Der Micrococcus der Variola

---

advice of a scientific committee, which at present consists of the following members: — John Simon, C.B., F.R.S.; John Tyndall, F.R.S.; John Burdon Sanderson, M.D., F.R.S.; and George Buchanan, M.D., F.R.S.

All communications on the subject are to be addressed to the clerk of the Grocers' Company, Grocers' Hall, London, E.C.

und Vaccine," Archiv f. experim. Pathol., Bd. X., Hft. 3 and 4, also Bd. 13, p. 284.

MARCHAND: "Les virus-vaccins." Revue Mycologique, IV. Année, Avril, 1882.

POHL-PINCUS: "Vaccination." Berlin, 1882.

SANDERSON (BURDON): Report on the Intimate Pathology of Contagion.

STRAUSS: "Vaccinal Micrococci." Société de Biol. de Paris. 1882.

STROPP: "Vaccination u. Mikrokokken." 1874.

TSCHAMER: "Ueber das Wesen des Contagiums der Variola, Vaccine, und Varicella." Aerztl. Verein, Steiermark, 1880, Sept. 8.

WEIGERT: "Ueber Bakterien in der Pockenhaut." Ctbl. f. d. med. Wiss., 1871, No. 49.

INFLUENZA EPIDEMICA.

Micrococcus influenzæ.

LITERATURE.

LETZERICH: Morpho. Unterschiede einigen path. Schizomyceten, Archiv f. experim. Pathol., Bd. XII., Hft. 5, p. 351 (1880).

MENINGITIS.

Micrococci.

LITERATURE.

AUFRECHT: "Zwei Fälle von Meningitis cerebrospinalis" (Mikrokokken). Deutsche, med. Wochenschr., 1880, No. 4.

GAUTIER : "Micrococcen bei Meningitis," Gaz. med. de Paris, 1881, Vol. XXXVII., 7.

LEYDEN : "Die Mikrokokken der cerebrospinalmeningitis." Ctbl. f. klin. Med., 1883, No. 10.

MEASLES.
Micrococci.

LITERATURE.

KEATING (J. M.) : "The presence of micrococcus in the blood of malignant Measles." Phila. Med. Times, No. 384.

KLEBS: I. "Der Micrococcus der Masern," Sitzungsber. d. phys. med. Gesellch. zu Würzburg, für 1873, p. VII. II. "Micrococcen als Krankheitursache," Berl. klin. Wochenschr., 1873, p. 116.

PHLEGMON.
Diplococci.

LITERATURE.

CORNIL : " Sur l'anatomie pathologique du phlegmon et, en particulier, sur le siége des bacteries dans cette affection." Compt. Rend. de l'Acad. de sc. de Paris, 1883, T. 97, p. 1594.

SCARLET FEVER.
LITERATURE.

CAZE U. FELTZ : "Malad. Infect." 1872.

HAHN (E.) :" Demonstration des Scharlachs auftretenden Schizomyceten," Berl. med. Gesellsch,

Marz, 1883 ; Berl. klin. Wochenschr., No 38, 1882, p. 583.

KLEIN (E.) : "Report of the Medical Officer of the Privy Council for 1876."

McKENDRICK : Brit. Med. Journ., 1872.

POHL-PINCUS : "Mikrokokken in dem Epidermisschuppen von Scharlochkranken." Ctbl. f. d. med. Wiss., 1883, No. 36.

MOLLUSCUM CONTAGIOSUM.

LITERATURE.

ANGELAUS : " Zur aetiologie von Molluscum contagiosum," Abth. f. med. Wiss., 1881, 3.

DOMENICO (M.) : " Sul Bacillo del mollusco contagioso." Atti. Acad. Lincei, Transunti. V., 1880, p. 77. — Gazetta Medica di Roma., 19, 20, 1880.

DILATATION AND OTHER DISEASES OF THE STOMACH.

Sarcina ventriculi (Goodsir).

LITERATURE.

EBERTH : " Ueber Sarcina ventriculi." Virchow's Archiv, 1858, Bd. XIII., p. 522.

KLEBS : " Ueber infectiose Magenaffectionen." Allgem. Wiener med. Zeitung, 1881, No. 29.35.

LOSDORFER : " Ueber Sarcina ventriculi." Med. Jahrb., 1871, H. 3.

SURINGAR : "De Sarcina ventriculi," Arch. Neerland, 1866 ; Botan. Zeitung, 1866, p. 269.

Virchow and Cohnheim : "Ueber Sarcina ventriculi," Virchow's Archiv, Bd. 10 and 33.

AMAEMIA PERNICIOSA.
(*Micrococci*).

Frankenhauser : Ctbl. f. d. med. Wiss. No. 4, 1883.

ULCERATIVE STOMATITIS IN THE CALF.
(*Bacilli.*)

Lingard (A.) and Batt (E.) : Lancet, May, 1883.

CATTLE PLAGUE.
(*Bacilli.*)

Archangelski : Ctbl. f. d. med. Wiss. No. 18, 1883.

Lanzi : "Microbe di Morbillo." Bull. Accad. Med. Roma. IX., 1883, No. 7.

Metzdorf : (Bacillus of Cattle Plague). Bied. Centr., 1884, p. 419, 420. Journ. Chem. Soc. (Abstr.), xlvi. (1884), p. 1398.

Roll : "Die Thierseuchen." Wien, 1881.

Roloff (T.) : Archiv f. wiss. u. prakt. Thierheilkunde., IX., 1883.

ACUTE YELLOW ATROPHY OF THE LIVER.
(*Micrococci.*)

Eppinger : Prager Viertelj. 1875.

Balzer (F.) : " Parasitisme du Xanthelasme

et de l'ictère grave." Archiv d. phys. nor. et. path. 1882, X., p. 307.

DYSENTERY.

(*Micrococci and Bacilli.*)

PRIOR: "Die Mikrokokken bei der Dysenterie." Ctbl. f. klin. Med. 1883, No. 17.

LINGARD (A.): Klein's "Micro-organisms and Disease." 1885, p. 128.

OZÆNA.

(*Diplococcus.*)

LOWENBERG: "Die Natur und die Behandlung der Ozæna." Union Med. No. 22–25, 1884. Deutsch. med. Wochenschr. 1885, 1, 2.

FOOT ROT.

TOUSSAINT: "Sur la culture du microbe de la clavelée." Compt. Rend. XCII., 1881, p. 362–4.

VERRUGA PERUANA.

(*Bacilli.*)

IZQUIERDO (V.): "Spaltpilze bei der Verruga peruana." Virchow's Archiv, Bd. 99, Hft. 3, p. 411.

DIABETES.

FRIEDRICH (N.): "Ueber das constante Vorkommen von Pilzen bei Diabetischen." Virchow's Arch., vol. XXX., p. 476, 1864.

CHYLURIA.
(*Bacilli.*)

WILSON (A.) : "A case of chyluria caused by Bacilli, with cultivation experiments." Brit. Med. Journ., 1884, Dec. 6.

FŒTID FEET.

THIN (G.) : "*Bacterium fœtidum:* an organism associated with profuse sweating of the soles of the feet, cultivated in vitreous humor." Proc. Roy. Soc., XXX. (1880), p. 473.

MALLEUS HUMIDUS.

DIRNER : "Ein an einem Menschen beobachten Fall von Malleus humidus."

ROZAHEGGI : "Bakterien im Eiterpusteln bei Malleus humidus." Pestes med. chir. Presse. 1883, No. 35.

PYÆMIA.

BELZOW (A.) : "Zur Frage der Mikroorganismen bei Pyæmia." Ctbl. f. d. med. Wiss. 1884, No. 22, p. 370.

KOCH (R.) : "Untersuchungen über die Aetiologie d. Wundinfections-Krankheiten. Leipzig, 1878.

DISEASES OF INSECTS.
LITERATURE.

BECHAMP : "Microzyma bombycis. Compt. Rend. T. 64, 1867.

BURRILL (T. J.): Disease of insect *Blissus leucopterus*. Amer. Naturalist, XVII., 1883, p. 319, 320. Named the microbe "*Micrococcus insectorum*."

FORBES (S. A.): "Bacterial Parasite of the chinch-bug (*Micropus leucopterus*)." Amer. Naturalist, XVI., 1882, pp. 824. II. "Experiments with diseased caterpillars (Schlaffsucht), claims the microbe found in *Datana ministra* is identical with *Micrococcus bombycis*. He also found *Micrococci* in *Datana angusii* and in the cabbage-worm, *Pieris rapæ*.

PASTEUR (L.): "Etude sur la maladie des vers à soie, moyen pratique assuré de la combattre et d'en prévenir la retour." 1870, I., p. 228.

BACTERIA IN PLANT TISSUES.

Ducleaux studied the effect of microbes upon germination by planting peas and beans in soil previously sterilized and moistened with milk also sterilized. There was no germination, and at the end of two months the milk showed no indication of alteration, the whole apparatus in which the experiment was conducted being, of course, protected against contamination from the air.

Ralph's Method for demonstrating the presence of *Bacilli* in the cells of water-plants (*Vallisneria*, etc.). A thin section of the cuticle of the leaf should be sliced off and placed on a slide, with the cuticular surface next the cover; the slide should then be placed on a rest, with the cover downwards or towards the table, and allowed to remain there for five minutes at least, to allow the organisms to fall on the cuticular walls of the cells, then examine under a quarter-inch objective. The Bacteria must be looked for in the quadrate cells, where they may be seen moving about the chlorophyll-grains even when cyclosis is going on. After the lapse of some minutes they will gravitate out of sight, or be found heaped together at the lower end (or apparent upper end) of the cell. It is this circumstance which has prevented any recognition of their presence heretofore; they are rarely if ever seen in the long, deep-seated cells which exhibit cyclosis so well in *Vallisneria*.

LITERATURE.

BATILIN: "Bacterien inficirte Weizen und Maiskörner." Botan. Zeitung, 1882, p. 28.

BURRILL (T. J.): "Micrococcus toxicatus, the peculiar poison of *Rhus*." Amer. Naturalist, XVII., 1883, p. 319. II. "Disease of Roots of Strawberry Plants." *Bacillus* shown at regular meeting of State Microscopical Society of Illinois, Feb. 8, 1884.

DUCLAUX: "Effect of presence of microbes upon germination." Meeting of Acad. of Sci., Paris, Jan. 5, 1885.

RALPH (T. S.) : "Living Bacilli in the cells of *Vallisneria* and *Anacharis*." Journal of Microscopy, iii. (1884), p. 17. — Proc. Roy. Soc. Victoria, 10 May, 1883.

REINKE U. BERTHOLD : (Plant pathology). Untersuchungen aus dem botan. Laboratorium in Gottingen. Hft. I.

WALKER : "Yellow disease of Hyacinths." Botan. Centralblatt. Bd. 14, p. 315.

BACTERIA INVESTIGATION.

PART III.

FORMULARY.

1. BERGMANN'S OR BUCHOLTZ'S FLUID.

Distilled water	100	ccm.
White rock candy	10	"
Tartrate of ammonia	1	"
Phosphate of lime	0.5	"

2. BISMARCK-BROWN.

This takes the lead of all aniline dyes, in that it is not so easily washed out. The preparations may also be mounted in glycerine, which is not the case with other aniline dyes. It is also especially adapted to photography. Weigert makes a saturated solution with boiling distilled water, or one in which there is very little alcohol; after cooling, filter, set aside, and allow to settle. After staining, wash the sections for a short time in absolute alcohol. Whole pieces may be stained at once before cutting by using an alcoholic solution of Bismarck-brown. Baumgarten used a concentrated (one per cent.) acetic acid solution.

3. Glycerine Aniline-Brown.

Used by Koch for photographing. Make a concentrated solution of aniline-brown in equal parts of glycerine and water. Filter from time to time.

4. Aniline Oil.

This is a mixture of two similar aniline bases — toluidin and pseudotoluidin. It is a yellowish, oily fluid, whose saturated, watery solution allows of the dissolving of more coloring matter than the dilute solution of caustic potash first used by Koch. When it is used in connection with the aniline dyes, the latter may be added directly to it, and the two shaken together, the mixture being allowed to stand and steep in an oven at a temperature of about 95° F. for from fifteen minutes to twenty-four hours before using, or a concentrated alcoholic solution of the coloring material is poured into a ten per cent aniline-oil-water. The colors best adapted for use in connection with aniline oil are fuchsin, methyl-violet and gentian-violet. These solutions, especially at a temperature of 95° F., will, in from two to twenty-four hours, stain those Bacteria which, as a rule, show such stubborn resistance to staining fluids.

5. ANILINE WATER.

To make an aniline water that is permanent and does not need to be filtered, take —

Aniline oil or Toluidin . . . . 3 ccm.
Alcohol . . . . 7 "
Dissolve and add water . . . . . . 90 "

(Fraenkel.)

6. ANILINE YELLOW.

Aniline yellow . . . . 0.2 ccm.
Distilled water . . . . 10 "

7. ACIDULATED FLUIDS FOR DECOLORIZING.

Ehrlich and Balmer-Fraenzel used a (33 per cent) watery solution of nitric acid.

Friedländer used nitric acid and alcohol (33$\frac{1}{3}$ per cent).

Glacial acetic acid is very well adapted for clearing up sections in which the Bacteria are to be studied without staining, rendering them very distinct.

Hydrochloric acid, one part, to 70 per cent alcohol, one hundred parts. Hydrochloric acid, 20 ccm.; 90 per cent alcohol, 100 ccm.; water, 20 ccm.

Coze and Simon recommend the following mixture (*Brun's*).

Nitric acid . . . . 5 grammes.
Acetic acid . . . . 10 "
Distilled water . . . . 55 "

8. Ether.

In tissues containing much fat, this may be removed by the use of ether in connection with absolute alcohol. Allow the preparation to lie in the alcohol until it is entirely dehydrated, then place in the ether. If it causes the ether to become turbid, it has not been sufficiently long in the alcohol. After they have remained for some time in the ether (or chloroform) place again in alcohol, and then in water.

9. Alcohol.

All alcohol used in the treatment of Bacteria, especially in connection with staining, must be perfectly pure and without any acid reaction.

10. Canada Balsam.

It is important, for mounting stained Bacteria, to use Canada balsam free from chloroform, as Orth and others testify that it removes the color from the microbes.

Baumgarten recommends equal parts of Canada balsam (free from chloroform) and oil of cloves.

11. Chrysoidin.

Make a concentrated alcoholic solution. For use dilute this with one half the quantity of water.

12. CEMENT FOR GLYCERINE MOUNTS.

Pure Venetian turpentine is to be poured into some melted wax upon a water bath until a portion taken out on a glass rod becomes at once stiff and does not stick to the hand. (*Dr. Csokor, Wien.*)

13. CULTURE FLUIDS.
(vid. F. 31. 46. 1.)

(1) Liebig's extract of meat . . . . . . 1 part
    Water . . . . 100 "
(2) Gelatine . . . . 3 parts
    Phosphate of ammonia . . . . . . .25 "
    Water . . . . 100 "
(3) Meat extract . . . . 1 part
    Sugar . . . . 3 "
    Water . . . . 100 "
(4) *Pasteur's Fluid.*
    Distilled water . . . . 100 parts
    Cane sugar (pure) . . . . . . 10 "
    Ammonium tartrate . . . . . . 1 "
    Ash of yeast . . . . 1 "
(5) *Cohn's Fluid.*
    Distilled water . . . . 100 ccm. or 200 parts
    Ammonium tartrate . . 1 gramme   20 "
    Potassium phosphate . . 0.5 "   20 "
    Magnesium sulph. cryst. 0.5 "   10 "
    Calcium phosphate (tri-
        basic)   0.5 "   0.1 "

Fluids (4) (5) are not suitable for the cultivation of pathogenic organisms. In making broths, select as fresh meat as possible, allow half an hour's boiling for each pound, and calculate to have sufficient water to yield, ultimately, one pint of broth to each pound. When boiled, allow to cool, skim off fat, and neutralize with carbonate of sodium, filter through a sterilized filter into sterilized flasks. If the broth is not clear after once filtering, repeat; and, if upon standing, a sediment appears, decant, or boil with the white of an egg, and filter.

14. EOSIN-HÆMATOXYLIN.

(1) Eosin 5
    Distilled water 100
    Alum 2.5
    Glycerine 100.0
(2) Hæmatoxylin 5
    Absolute alcohol 100

Mix (1) and (2), After three days, during which the mixture is exposed to the light, add glacial acetic acid 2 per cent. (*Niesser.*)

15. EOSIN.

This dye appears in commerce as eosinate of potash. In a saturated solution of the salt a pure eosinic acid will be precipitated, if an acid is added. If the precipitate is collected upon a filter, and dissolved in strong alcohol it gives a solution of pure eosinic acid.

The ordinary eosinate of potash is used in a 5

or 10 per cent aqueous solution, of which one adds some drops to the fluid containing the sections. It is better to mount sections stained in eosin in glycerine containing alum.

16. FUCHSIN.

Used by Rindfleish in a concentrated alcoholic solution.

Used by Lichtheim in a concentrated aqueous solution.

Used by Pfuhl-Petri in following solution: —

    Saturated alcoholic solution of Fuchsin   10 ccm.
    Distilled water . . . . 100 ccm.

Burrill's Solution.

    Glycerine . . . . 20 parts
    Fuchsin . . . . 3 "
    Aniline oil . . . . 2 "
    Carbolic acid . . . . 2 "

A slight variation in the quantities does not matter, but the glycerine should not be decreased. Mix and shake carefully. Prepare a quantity of the solution in advance, as it does not deteriorate as do alcoholic solutions.

Gradle's Solution, recommended by Hartzell.

    Carbolic acid . . . . 15 minims
    Distilled water . . . . $\frac{1}{2}$ fluid oz.

Dissolve, and add alcoholic solution of fuchsin, $\frac{1}{2}$ fluid drachm.

17. FUCHSIN-ANILINE-OIL.

Add 10 or 20 drops of a concentrated alcoholic solution of fuchsin to 10 ccm. of aniline-water. Used by Ehrlich for tubercle *Bacilli* and by Baumgarten for lepra *Bacilli*.

18. GENTIAN-VIOLET.

(a) *Weigert's solution.*

Distilled water . . . . 90 grams
Absolute alcohol . . . . 10 "
Caustic ammonia . . . . . . 0.5 "
Gentian violet . . . . 2. "
    Mix, and filter.

(b) *Cose and Simon's Solution.*

Gentian-violet (or fuchsin) . . . . 2 grams
Alcohol (90 per cent) . . . . 5 "
Saturated aniline-water . . . . 100 "

Käätzers solution.

Make a saturated solution with 90 per cent alcohol.

19. ANILINE-GENTIAN-VIOLET.

(a) *Balmer-Fraenzel's solution*

Pure aniline oil, 1 ccm. } = Aniline water
Water . . . 24 ccm. }

Shake together, and pass through a wet filter-paper. To the clear filtrate add one half gram of finely powdered gentian-violet. Dissolve by stirring with a stick. Filter just before using. (2 grams to 100 ccm.)

(b) *Ehrlich's solution of Gentian-violet or Fuchsin.*

To saturated aniline water 10 ccm., add concentrated alcoholic solution of either the above (10–20 drops) until opalescence occurs (showing saturation). Gentian-violet is also used in concentrated alcoholic solutions; also in concentrated aqueous solutions (*Lichtheim*). Käätzer uses 10 drops of a saturated alcoholic (90 per cent) solution of gentian-violet to 10 cc. aniline water. Shake well, and filter. Gentian-violet gives a good double-staining with vesuvin.

20. KLEB'S GLYCERINE-JELLY FOR MOUNTING.

To be used warm, or a piece laid on the slide and this warmed. All air-bubbles must be driven off. Its advantages are that it hardens in cooling, so that a slide may be cleaned and finished at once.

Take best clear gelatine, 10 grams. Allow this to swell up in distilled water; throw away the water left, and dissolve the swollen glue by gentle warming. Then add 10 grams of glycerine and a few drops of phenol, to prevent the formation of mould.

21. GLYCERINE.

Use pure glycerine, entirely free from acid reaction. Glycerine is not adapted for mounting

Bacteria which have been stained in any of the aniline dyes, with the exception of Bismarck-brown. Microbes stained in Bismarck-brown and mounted in glycerine make the best preparations for photographing.

22. Boehmer's Hæmatoxylin.

(1) Dissolve 0.35 parts of crystallized hæmatoxylin in ten parts of absolute alcohol.

(2) Dissolve 1 part of alum in 30 parts of distilled water. Add the first solution gradually to the second, until a beautiful violet is developed.

Cose and Simon recommend the following solution. Make a 2 per cent hydro-alcoholic solution of hæmatoxylin and a 3 per cent solution of ammonia alum. A few drops of the hæmatoxylin added to a small quantity of the ammonia alum solution gives in the course of a few minutes a beautiful violet color.

23. Methyl-blue.

(a) *Koch's Solution.*

Take distilled water 200 cctm.

Concentrated alcoholic solution of Methyl-blue 10. Shake well, and add —

Ten per cent solution of caustic potash 0.2. Allow the mixture to stand for a day. It should give no sediment. Filter.

(b) *Fränkel's Solution*, For double-staining.

Alcohol 50.0
Water 30.0
Nitric acid 20.0

Saturate with methyl-blue, and filter. This solution removes from the matrix in which the Bacteria lie the dye first used, and replaces it with blue.

(c) *Schutz's Solution.*

Caustic potash solution (1–10,000,
Absolute alcohol, } Equal parts
Solution of methyl-blue,

(d) *Watson Cheyne's Solution.*

Distilled water . . . . 100 ccm.
Saturated alcoholic solution of methyl-
    blue . . . . 20 ccm.
Formic acid . . . . 10 minims
    (Practitioner, April, 1883, p. 258.)

24. METHYL-GREEN AND MALACHIT-GREEN.

(a) Of the former, use saturated aqueous solution (*Subbotin*).

(b) Of the latter, use a saturated alcoholic solution (*Pfuhl and Petri*).

25. METHYL-VIOLET.

Used by Baumgarten and Fraenkel in a saturated alcoholic solution.

Koch's Formula.

Take of pure aniline 5 cc., shake repeatedly with distilled water. In half an hour 3 or 4 per cent of the aniline is dissolved, the rest settles to the bottom. This mixture is filtered, and, if not absolutely clear, must be filtered again. Prepare new each time.

To 100–150 cc. of absolute alcohol add 20 grams of dry methyl-violet, allow to stand several days, shake repeatedly. Fuchsin may be used instead.

11 cc. of the alcoholic solution of methyl-violet is mixed with 100 cc. of the aniline-water, and the staining fluid is ready for use. Add a crystal of thymol to keep it, and filter before use.

26. MAGENTA.

Magenta crystals	2 grams
Pure aniline oil	3 ccm.
Alcohol (sp. gr. 0.830)	20 "
Water	20 "

Pulverize the crystals, dissolve in the aniline-oil and alcohol, and lastly add the water (*Gibbs*).

27. NIGROSIN OR ANILINE-BLACK.

Use a strong aqueous solution; wash sections in alcohol, and mount either in glycerine or Canada balsam (*Errera*).

28. ORSEILLE.

Orseille extract is derived from *Rocella tinctoria.*

Absolute Alcohol . . . . 20 cctm.
Glacial acetic acid (sp. gr. 1.070) . . 5 "
Distilled water . . . . 40 "

Add as much orseille extract as will form a saturated dark red fluid, which must be filtered two or three times, until a clear ruby fluid is obtained. This makes a good stain for the tissue containing Bacteria, as it stains cell-substance but not nuclei. Mount in levulose (*vid.* Virchow's Archiv, Bd. 71.) (*Wedl.*)

29. OIL OF CLOVES.

While this is most commonly used for clearing stained tissues, it may be replaced by other ethereal oils, *e. g.*, cedar, origanum, cinnamon, bergamot, lavender, anise, or by xylol or creosote.

30. OSMIC ACID.

This is a volatile, strong-smelling crystalline substance, very irritating to mucous membranes. The breathing of the fumes must be positively avoided. In commerce it comes in small, sealed glass tubes, each holding a gram or a half gram of the greenish, crystallized acid. It is used as a fixing and hardening medium in aqueous solutions of 1 to 500, or 1 to 100. The tube should be broken inside a bottle containing the proper quantity of *distilled* water, by means of a glass rod. The solution must be kept from the light, or it will be rapidly reduced.

31. Peptone Solution.

Peptone (chemically pure)	5.9
Bi-sodic phosphate	10.
Lactate of ammonia	5.
Liebig's beef extract	5.
Sugar	20.
Distilled water	1000.

Peptone, being a changeable product, is, if not fresh, best prepared especially for the purpose, *e. g.*, allow 500 grams of lean beef, 2 litres of water, and 1 gram of salt to simmer for four hours, in an uncovered vessel. Then close it, and heat for one hour to 100° C. Cool, and skim off the fat; neutralize carefully, and filter.

Buchner's Fluid.

Leibig's extract	10 parts
Peptone	8 "
Water	1000 "

32. Picrocarmine.

Add aqua ammonia, 4 grams, to carmine, 2 grams, and allow it to stand for twenty-four hours in a damp place, and then add 200 grams of picric acid. Allow the whole to remain for twenty-four hours longer, until all is dissolved that will dissolve. Filter, and to the filtrate add a small quantity of acetic acid, until it becomes turbid. After twenty-four hours more there is a precipitate, and the filtered fluid also remains turbid. Now add ammonia, drop by drop, and allow the solution each time to remain for twenty-four hours, until at length, in the course of a few days, it remains entirely clear. If the neutral solution stains too yellow, add a little acetic acid; if too red, a little ammonia. (*Weigert.*)

33. Picrocarminate of Ammonia.

Pour into a saturated solution of picric acid a strong ammoniacal carmine solution until it becomes turbid. Evaporate this mixture to one fifth of its original volume; filter. Now evaporate the filtrate to dryness, and add the red powder resulting each time, before use, to about one hundred parts of water, or allow the original solution to dry in the open air, in an evaporating dish; dissolve the residue in a little water, filter, and add to the filtrate some carbolic acid, to prevent fermentation. (*Ranvier.*)

34. Acetate of Potash Mounting Fluid.

This is simply a strongly concentrated solution of acetate of potash. It behaves much as glycerine, does not dry at the edges, and is less refractive. (Refractive index for the line D, 1, 370. *Ch. Soret.*)

Many aniline-stained preparations of Bacteria keep well in this solution. (*Max Schultze.*)

35. Caustic Potash Solutions.

(*a.*) Caustic potash fus. . . . . . 1 gram
Absolute alcohol . . . . . . 100 ccm

Allow to stand twenty-four hours, until the alcohol is saturated. Decant, and mix with ten times the volume of pure alcohol.

(*b.*) Caustic potash fus. . . . . . 20 parts
Water 100

(*c.*) Baumgarten uses carbonate of potash in his method, in a half saturated solution for decolorizing.

(*d.*) *Iodo-Iodide of Potash*, solution for decolorizing.
Iodine 1.0
Iodide of potash 2.0
Water 300.0

36. Rosaniline-chlorhydrate.

Rub up 2 grams of rosanilin-chlorhydrate with 1 gram of methyl-blue, and dissolve slowly in 15 ccm. of alcohol. To this add 3 ccm. of aniline oil and 15 ccm. of distilled water. Keep in a stoppered bottle. (*Gibbes.*)

37. Acid Fuchsin (Rosanilinsaures Natron).

A saturated aqueous solution is used.

38. VESUVIN.

This combines well for double staining with methyl-blue or gentian-violet.

Fraenkel's Solution, for double staining.

Alcohol	70.0
Nitric acid	30.0

Saturate with vesuvin, and filter.

39. PLASTER OF PARIS MIXTURE FOR FILTER.

Water	46.
Plaster-of-Paris (such as is used for modelling)	52.4
Asbestos	1.6

Mix carefully, adding the plaster gradually.

40. CLEANING FLUIDS FOR SLIDES AND COVERS.

(*Hanaman's Formula.*) — To a cold saturated solution of bichromate of potash, add $\frac{1}{8}$ of its bulk of strong sulphuric acid, (care must be taken on account of the heat and vapors evolved). Journ. Roy. Mic. Soc. i. (1878) p. 295. Amer. Naturalist, XII. p. 573.

(*Gibbs's Method*).—Place the cover-glasses in strong sulphuric acid for an hour or two. Wash well, until the drainings give no acid reaction; wash first with methylated spirits, and then with absolute alcohol, and wipe carefully with an old silk handkerchief. Journ. Roy. Mic. Soc. iii. (1880) p. 392.

(*Seiler's Method.*) — New slides and covers are placed for a few hours in the following solution :—

Bichromate of potash . . . . 2 ounces
Sulphuric acid . . . . 3 fluid "
Water . . . . 25 " "

Wash with water, the slides may be simply drained dry; the covers may be wiped dry with a linen rag. Slides and covers that have been used for mounting either with balsam or a watery medium are treated as follows :—

The covers are pushed into a mixture of equal parts of alcohol and hydrochloric acid, and, after a few days, are put into the bicromate solution and treated like new ones. The slides are scraped free of the mounting medium and put directly into the bichromate solution. *Ibid.* p. 508.

41. Soap Imbedding Mass.

Take good white soap, cut it up into thin slices, and put them to dry in the sun for some days, until they become white. The slices are then to be rubbed up to a fine powder, which is mixed with spirit to the consistency of porridge. Now mix the porridge with alcohol and glycerine in such proportions that the whole shall contain, for every 10 parts by weight of the soap, 22 parts of glycerine, and 35 parts of alcohol, (90 per cent). Let the whole simmer until there is obtained a perfectly transparent, syrupy, somewhat yellow fluid. The objects, previously dehydrated in alcohol, are imbedded in this mass in the usual manner. The mass may be removed from the sections either by means of water, or of very dilute alcohol. It has the following advantages. 1. It is transparent. 2. It adapts itself perfectly to the objects. 3. It cuts remarkably well. (*Salensky.*)

42. CELLOIDIN IMBEDDING MASS.

Celloidin is a preparation of pure pyroxylin. It is manufactured by the Chemische Fabrik auf Actien (vorm. E. Schering), Berlin, N. Fenstrasse, 11. 12. It may be obtained by mail by writing to Schering's Grüne Apotheke, Wittick and Benkendorf, Berlin, N. Chaussée-Strasse, No. 19 [or from wholesale dealers in the United States]. It is stated to be prepared with the purest pyroxylin, and to be always of uniform composition. It is sent in the form of tablets of a tough, gelatinous consistency and slightly milky-white transparency. These tablets have exactly the consistency that is required for section-cutting. They contain 20 per cent of pure pyroxylin. Celloidin is soluble in all proportions in ether and alcohol. It is free from acids. Merkel and Schifferdecker speak of celloidin as an imbedding medium, as "performing more, less hurtful to the tissues, and easier to manage, than any other known imbedding mass." Cut the celloidin into small pieces, and dissolve in

BACTERIA INVESTIGATION. 257

equal parts of absolute alcohol and ether. The tissues to be imbedded are thoroughly soaked in absolute alcohol, from which they are brought into the celloidin (in a well-closed vessel), and remain there until thoroughly impregnated (from a few minutes to eight days or more, according to thickness). If the objects contain cavities that it is desired to fill, it is best to use a thinner and therefore more penetrating solution of celloidin. When soaked, remove the preparations to a paper tray (or simply a small piece of leather), surround them with celloidin; wait a few minutes, until a skin has formed on the celloidin, and throw them into alcohol of 82° Richter (a considerable quantity of alcohol should be taken). After twenty-four hours the preparations on leather are generally fit to be cut, whilst those in paper trays may have the paper removed and be put back in the alcohol for twenty-four hours more. The preparations may remain in the alcohol for any length of time without harm. Sections are cut with a knife moistened with common alcohol; they are floated into either water or alcohol. Without any further manipulation they may now be stained with the usual staining agents, just as if they were not imbedded. Dehydrate in 95 per cent alcohol (not absolute, or the celloidin will be dissolved) clear with oil of bergamot, sandalwood, or origanum, (the last by preference), and mount in balsam or castor oil. Oil of cloves should *not* be employed, because it dissolves the celloidin.

Sections may be cleared and mounted in glycerine, which suffices to make the celloidin as clear as glass.

Instead of hardening in alcohol, the tissues, after being thoroughly impregnated with celloidin, may be allowed to dry under a bell-glass, or may be brought into chloroform. Under the influence of this reagent, the celloidin coagulates into a mass having the consistence of wax, but having also an elasticity that renders it unbreakable, and having besides the precious quality of being admirably transparent, and possessing exactly the index of refraction of glass. (From *Microtomist's Vade Mecum.* See p. 197–199.)

43. Brun's Mounting Medium.

Glycerine	10 parts
Glucose of commerce	40 "
Spts. of camphor	10 "
Water	140 "

Mix and filter. The advantage of this medium is that it possesses the refractive index 1.37 in the yellow ray, an index corresponding to that of the substances composing sputum (albumen 1.36, saliva and mucus 1.34, pus 1.39, *Robin*). Ordinary Canada balsam has a very high index, 1.53; in which Bacteria are poorly defined; colorless oils, for example, that of rape or castor are preferable, but their index is also too high, 1.48.

44. Gautier's Red-Lead Cement.

Crystalized boracic acid . . . . . 8 parts
Silicic acid . . . . 2 "
Red lead . . . . 12 "

45. Miquel's Nutritive Paper.

Pour upon a large sheet of black or gray paper, boiling lichen jelly (F. 46a), and spread it out uniformly, so that in the wet state it has a thickness of 2–3 mm. This done, dry it rapidly in an oven, at 40°. In a few hours the sheet of paper is dry. It is thin, flexible, it never presents cracks, and resembles altogether the photographic paper covered with bromo-gelatine emulsion, invented by Dr. Maddox. This nutritive lichen jelly paper may be preserved indefinitely with all its qualities, it being only necessary to place it in a dry drawer. At the moment of making the experiment, the paper is sterilized by being suspended in an atmosphere of vapor at 110° C. This is accomplished without its running, and it is scarcely any swollen. Dr. Maddox proposes that collodion films be used instead of paper, on account of the greater transparency and absence of structure.

46. Solid Culture Media.

(a) *Miquel's Lichen Jelly.*

Digest in one litre of beef broth, 25 to 30 grams of Irish moss (*Chondrus crispus*), and pass the resulting decoction through a sieve which retains the swollen leaves. After neutralization and a short boiling, filter the broth through bolting-cloth. Upon cooling it forms a strong jelly. To obviate the loss of broth which remains in the swollen leaves, Miquel makes in his laboratory a lichen jelly with water, this he dries and adds to broth in proper proportions, about 1 per cent. This nutritive jelly possesses the following advantages. 1. It melts only between 55° and 60° C., which permits of the cultivation of such organisms as require for their development elevated temperatures. Ordinary nutritive gelatines melt before 30° C. 2. It remains without alteration or losing its power of solidifying when exposed to a temperature of 110° C., for rigorous sterilization. Gelatine, on the contrary, is reduced under such conditions to a turbid broth, which remains fluid on cooling. Dr. Miquel does not favor those micrographic methods, introduced from Germany, by which one is recommended to use an intermittent heat of 100° C. All germs are not destroyed by this degree of moist heat, and if one obtains gelatines having the appearance of being sterilized, it is for the simple reason that the germs, being buried in the gelatine and deprived of oxygen, are under bad conditions for giving a visible growth.

(b) *Hydrocele Fluid.* (Koch.)

Use, for tapping, a thoroughly sterilized trocar and canula, connected by a sterilized rubber tube with a flask also sterilized. The collected fluid is then in its turn sterilized, by exposure to a temperature of 58°–62° C., for three to five hours, for five or six consecutive days. This may be considered accomplished if the fluid remains limpid in the incubator at 32°–38° C., for several weeks. This limpid fluid may be used in fluid cultures alone, and in solid cultures mixed with gelatine, or it may be made solid in the same manner as blood serum by heating it *gradually* up to 60°–70° C. In the course of an hour it solidifies. It is important to proceed gradually in this process, as there is danger of its losing its limpidity, and the heating should be done in vessels in which it is to be used, as it cannot be again liquefied. A better method of rendering hydrocele fluid and blood serum solid through evaporation is to expose them, in properly protected sterilized test tubes, to a heat of 32°–38° C. for several weeks. If prepared in this way, the limpidity is not lost.

(c) *Blood Serum.* (Koch.)

Blood of a healthy sheep is drawn from the carotid artery, by means of a canula, into a flask, with which it is connected by a rubber tube, the whole apparatus being first thoroughly sterilized. After standing twenty-four hours, or until a firm clot has formed, the serum is drawn off by means of a sterilized glass siphon, one end of which passes through the cotton plug in the flask containing the blood, and the other through the plug in the vessel intended for the reception of the serum. The serum is then sterilized, and, if desired, rendered solid by the same process as that used for hydrocele fluid.

(d) *Agar Agar.*[1]

This substance occurs in commerce in bunches of thin, shrivelled, transparent strips. It is prepared for use by soaking over night in salt water (1–6), and then dissolved by the aid of heat. After being filtered and neutralized, it is mixed with some nutritive material. Klein says that among all solid media he finds a mixture of agar agar and peptone the best. He prepares it by mixing peptone with the filtered agar agar solution, boils it repeatedly for thirty minutes at a

---

[1] To be had of Dr. Georg Grübler, 17 Durfour Strasse, Leipzig; Physiologische-chemische Laboratorium, or of Messrs. Christy & Co., 155 Fenchurch Street, London.

time, and finally obtains a sterile transparent mass, which remains solid up to 45°– 50° C. It becomes liquid at higher temperatures, and, in case of necessity, may be again subjected to boiling. Before considering it as perfectly sterile, it ought to be kept, like all other materials, for from several days to several weeks in the incubator, at 32°– 38° C. If quite limpid after this time it may be considered as sterile.

(e) *Gelatine.*

The best gelatine is cut up into strips and soaked in distilled water over night (1 part of gelatine to 6 of water), and is then melted, thoroughly neutralized, and filtered through sterilized bolting cloth. If it is not clear, it is boiled with some white of egg and again filtered. The fluid gelatine is then mixed with half its bulk of broth, peptone solution, or beef-extract solution, so that there is 1 part of gelatine in 9 parts of fluid, or 11½ per cent of gelatine. This mixture is boiled repeatedly, and treated like fluid culture media, as described above. It may be kept on hand as a stock, either with the addition of the peptone, etc., or as simple gelatine, which can be added to any particular nourishing material when desired.

www.ingramcontent.com/pod-product-compliance
Lightning Source LLC
Chambersburg PA
CBHW031947230426
43672CB00010B/2076